P. Bongrand   P. M. Claesson   A. S. G. Curtis   (Eds.)

Studying
Cell Adhesion

With 95 Figures

Springer-Verlag
Berlin Heidelberg New York London Paris
Tokyo Hong Kong Barcelona Budapest

Professor Dr. PIERRE BONGRAND

Hopital de Sainte-Marguerite
Laboratoire d'Immunologie
INSERM 0387
13277 Marseille Cedex 09
France

Dr. PER M. CLAESSON

The Surface Force Group
Department of Physical Chemistry
The Royal Institute of Technology
S-114 86 Stockholm
Sweden

Professor Dr. ADAM S. G. CURTIS

Department of Cell Biology
University of Glasgow
Glasgow G12 8QQ
Scotland, U.K.

QH
623
,S78
1994

ISBN 3-540-57590-1 Springer-Verlag Berlin Heidelberg New York
ISBN 0-387-57590-1 Springer-Verlag New York Berlin Heidelberg

Libarary of Congress Cataloging-in-Publication Data. Studying cell adhesion / P. Bongrand, P.M. Claesson, A.S.G. Curtis, eds. p. cm. Includes bibliographics references. ISBN 3-540-57590-1. ISBN 0-387-57590-1 (New York)   1. Cell adhesion - Research - Methodology. I. Bongrand, Pierre. II. Claesson, P.M. (Per M.) III. Curtis, A.S.G.   QH623.S78 1994   574.87'6-dc20   94-36098

© Springer-Verlag Berlin Heidelberg 1994
Printed in Germany

Cover design: Struve & Partner, Heidelberg
Typesetting: Camera ready by author
39/3130-5 4 3 2 1 0 - Printed on acid-free paper

Two Preliminary Remarks

P.G. de Gennes

We know that cellular adhesion plays an essential role in tissue build up and tissue function. But it is hard to measure. Indeed, it is even hard to define.

When we join two *solid* blocks via a glue, our first (naive) mode of thinking focuses on the critical stress σ_c required to separate them. However, this is *not* the material parameter defining the quality of the glue. The true material parameter is the *adhesion energy* G, i.e. the energy per unit area required to separate the two partners. Of course, the critical stress σ_c is related to G, but it also depends on the thickness W of the glue layer : $\sigma_c \sim (\mu G/W)^{1/2}$, where μ is the elastic modulus of the glue. Thus σ_c is not a material parameter.

Cell / cell adhesion is a case of bonding between *soft* objects. But, here also, it is not well expressed in terms of a separation force ; it is defined more fundamentally by an adhesion energy G. Delicate micromanipulations on cell pairs sometimes allow to measure G (see for instance the paper by E. Evans and coworkers in this book). Figure 1 shows another type of set up, which may soon become available : here, an optical tweezer allows one to pull out a cell from a contact surface.
The pulling force F induces a tension τ in the cell :

$$F = 2\pi r \tau \sin \Theta \tag{1}$$

and the force balance on the contact line gives :

$$G = \tau(1 - \cos \Theta) \tag{2}$$

Knowing F and Θ we should, in principle, be able to measure G.

Needless to say, most adhesion measurements on living cells measure only a certain threshold force, similar to the F described above. And this is sufficient for many purposes, e.g. if we want to classify the adhesion properties of one cell type under different effectors. The present book is

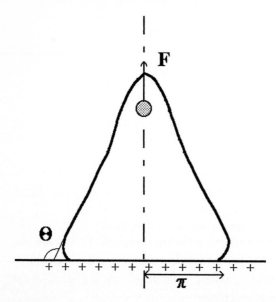

Fig.1. Separation of a cell from a contact surface with an optical tweezer. The case shown here corresponds to a very permeable cell, with an adjustable water volume.

mainly dedicated to this type of measurements : certain set ups, allowing for a fast collection of data, are extremely ingenious.

For future, quantitative studies, the definition of adhesion in terms of the energy G will have to be kept in mind. But another warning is also essential. When we look at certain classical patterns of embryological development, we are often tempted to relate them to interfacial energies (γ) between cells. Indeed, in certain reconstitution experiments, a mixture of two cell types A and B segregates very much like two non compatible fluids, and this may possibly be related to the difference

$$\gamma_{AB} - 1/2(\gamma_{AA} + \gamma_{BB})$$

It must be stressed that there is, in general, *no simple relation* between the adhesion energy G and the interfacial energy γ. Many forms of irreversible dissipation contribute to G, and not to γ.

A simple example of this is a "connector" (Fig. 2) firmly bound at both ends to the cytoskeleton of the two partner cells.

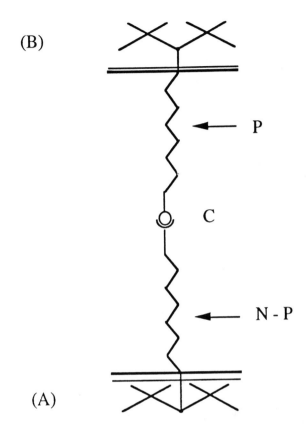

(B)

P

C

N - P

(A)

Fig.2. A "connector" relating cell A to cell B.

The connector is made up of one recognition site C plus two spacers. The spacers are assumed to be more or less linear (not globular) : they are formed of a total of N repetitive units (each of them can be either flexible or rigid : this is not the problem). When the two complementary pieces, emerging from both cells, establish contact at point C, this implies a certain binding energy U_b. The equilibrium interfacial energy is $\gamma = \nu\, U_b$, where ν is the number of connectors per unit area. But the adhesion energy G is *much larger* :

$$G \sim \nu\, N\, U_b \tag{3}$$

This was first understood, in a different context, by Lake and Thomas. The point which they made may be stated as follows : when I hold an elastic rubber band under tension, if somebody comes with a pair of scissors and cuts it into two halves, the two halves snap back on my hands, and the energy which I (painfully) experience on my hands is much larger than the energy required for the scissors to clip the band.

Thus, returning to the very first sentence of this foreword, tissue build up may (sometimes) be understood in terms of γ, but tissue adhesion is described by G, and the two things are distinct.

It is important to remember these two caveats. Of course, the measurements on cellular adhesion which are currently performed require many other, delicate precautions, related to cell preparation, to the possibility of spurious damage during separation, and also to the intrinsic complications brought in by poorly shaped objects.

The present book is remarkably well tuned to these problems. The editors combine three forms of knowledge : physico-chemical, biological, medical. The spectrum of authors is wide, but the langage is simplified, and hopefully accessible to all. I am convinced that this joint production will have long lasting success.

P.G.de Gennes
March 1994

Contents

List of Contributors

François Amblard, Department of Molecular Biology, Princeton
University, Princeton, NJ 08544, USA

Anne-Marie Benoliel, INSERM U 387, Laboratoire d'Immunologie,
Hôpital de Sainte-Marguerite, BP 29, 13277 Marseille Cedex 09, France

Pierre Bongrand, INSERM U 387, Laboratoire d'Immunologie,
Hôpital de Sainte-Marguerite, BP 29, 13277 Marseille Cedex 09, France

Christian Capo, INSERM U 387, Laboratoire d'Immunologie,
Hôpital de Sainte-Marguerite, BP 29, 13277 Marseille Cedex 09, France

Shivkumar Chiruvolu, Department of Chemical Engineering and
Materials Department, University of California, Santa Barbara, California
93106, USA

Per M. Claesson, The Surface Force Group, Department of Physical
Chemistry, The Royal Institute of Technology, S-114 86 Stockholm,
Sweden

Adam Curtis, Department of Cell Biology, University of Glasgow,
Glasgow G12 8QQ, Scotland, U. K.

Ragnar Erlandsson, Department of Physics and Measurement
Technology, Linköping University, S-581 83 Linköping, Sweden

Evan Evans, Departments of Pathology and Physics,
University of British Columbia, Vancouver, B.C., Canada V6T 1W5

Colette Foa, INSERM U 387, Laboratoire d'Immunologie,
Hôpital de Sainte-Marguerite, BP 29, 13277 Marseille Cedex 09, France

Marc Fraterno, Service commun de Microscopie Electronique, Faculté de Médecine, 27 Bd Jean Moulin, 13005 Marseille, France

David Gingell, Department of Anatomy and Developmental Biology, University College, Gower Street, London WC1E 6BT, Great Britain

Harry L. Goldsmith, Department of Medicine, Hôpital Général de Montréal, 1650 Avenue Cedar, Montréal, Québec H3G 1A4, Canada

Philip G. de Groot, University Hospital Utrecht, Department of Haematology, G03.647, Postbox 85500, 3508 GA, Utrecht, The Netherlands

Jacob Israelachvili, Department of Chemical Engineering and Materials Department, University of California, Santa Barbara, California 93106, USA

Jean-Louis Lavergne, Ecole Centrale de Lyon, Laboratoire de Technologie des Surfaces, UA CNRS 855, 36 Avenue Guy de Collongue, BP 163, 69131 Ecully Cedex, France

Deborah Leckband, Department of Chemical Engineering and Materials Department, University of California, Santa Barbara, California 93106, USA

Jean-Michel Martin, Ecole Centrale de Lyon, Laboratoire de Technologie des Surfaces, UA CNRS 855, 36 Avenue Guy de Collongue, BP 163, 69131 Ecully Cedex, France

Jean-Louis Mège, INSERM U 387, Laboratoire d'Immunologie, Hôpital de Sainte-Marguerite, BP 29, 13277 Marseille Cedex 09, France

Rudi Merkel, Department of Physics, University of British Columbia, Vancouver, B.C., Canada V6T 1W5

Philippe Naquet, Centre d'Immunologie de Marseille-Luminy, Case 906, 13288 Marseille Cedex 09, France

Lars Olsson, Department of Physics and Measurement Technology, Linköping University, S-581 83 Linköping, Sweden

John L. Parker, The Surface Force Group, Department of Physical Chemistry, The Royal Institute of Technology, S-100 44 Stockholm, Sweden and
The Institute for Surface Chemistry, Box 5607, S-114 86 Stockholm, Sweden (On leave from the Departments of Applied Mathematics, Research School of Physical Sciences, G.P.O Box 4, Canberra A.C.T. 2601, Australia)

Marc Passerel, Centre de Microscopie Electronique et de Microanalyse, Faculté des Sciences et Techniques de Saint-Jérôme, Avenue Escadrille Normandie-Niemen (Case 151), 13397 Marseille Cedex 13, France

Anne Pierres, INSERM U 387, Laboratoire d'Immunologie, Hôpital de Sainte-Marguerite, BP 29, 13277 Marseille Cedex 09, France

Ken Ritchie, Department of Physics, University of British Columbia, Vancouver, B.C., Canada V6T 1W5

Franz-Josef Schmitt, Department of Chemical Engineering and Materials Department, University of California, Santa Barbara, California 93106, USA

Jan J. Sixma, University Hospital Utrecht, Department of Haematology, G03.647, Postbox 85500, 3508 GA, Utrecht, The Netherlands

Mireille Soler, INSERM U 387, Laboratoire d'Immunologie, Hôpital de Sainte-Marguerite, BP 29, 13277 Marseille Cedex 09, France

Koichi Takamura, Department of Medicine, Hôpital Général de Montréal, 1650 Avenue Cedar, Montréal, Québec H3G 1A4, Canada

David Tees, Department of Medicine, Hôpital Général de Montréal, 1650 Avenue Cedar, Montréal, Québec H3G 1A4, Canada

Susan Tha, Department of Pathology, University of British Columbia, Vancouver, B.C., Canada V6T 1W5

Olivier Tissot, INSERM U 387, Laboratoire d'Immunologie,
Hôpital de Sainte-Marguerite, BP 29, 13277 Marseille Cedex 09, France

Scott Walker, Department of Chemical Engineering and Materials
Department, University of California, Santa Barbara, California 93106,
USA

Joe Zasadzinski, Department of Chemical Engineering and Materials
Department, University of California, Santa Barbara, California 93106,
USA

Andreas Zilker, Research Division, Daimler-Benz, Ulm, Germany

1 Introduction

A. S. G. Curtis, P. Bongrand and P. Claesson

1.1 Why Study Adhesion ?

Cell adhesion, both of cells to cells and of cells to other objects, is a very common phenomenon though there are a few cells such as erythrocytes remarkable for their lack of adhesion in most circumstances. Adhesion and changes in adhesion form an important feature of the normal developmental processes of animals (Edelman and Thiery 1985), are often features of pathogenesis of disease (e.g. Harlan and Liu 1990), are needed for the replacement and repair of tissues, are essential for the fusion of gametes, and frequently play a major role in infectivity or parasitism (Ellwood et al 1978).

This book has been planned to describe general methodological approaches rather than recipes. This choice is based on the following considerations :

i) Due to the complexity of cell behaviour, the quantitative determination of primary parameters such as bond strength, binding affinity or cell-surface distance requires the use of sophisticated devices that may not be commercially available (see, e.g. chapters by H. Goldsmith, E. Evans, F. Amblard) and may require several months or more to mount. Therefore, it will be more useful for the reader to obtain information about the *potential* of these methods before starting a long-term project and/or collaboration than to obtain "recipes" that would be in any case insufficient to allow a newcomer to the field to practice a given method.

ii) Another point is that many methods are not fully standardised and may require some "custom" adaptation. It is therefore useful to compare different procedures used in different laboratories to study different experimental systems with possibly different biological significance and behaviour. Thus, two different ways of dealing with a flow chamber are described in the chapters by Tissot et al. and Sixma et al.

iii) Finally, a specific feature of many methods is that the significance of the measurements obtained is not always obvious and requires critical interpretation. Thus, the interpretation of many sets of data requires a fairly wide physical and biological background that is difficult to find in a specific textbook. Hence, the authors have included more introductory material of general interest than usually found in methods books. Also, a fairly extended glossary has been added .

1.2 The Need for Quantitation

Most scientists accept that measurement is an essential feature in the understanding of a mechanism. As you will see from this book, the accurate absolute measurement of adhesive forces or energies is often very difficult so that in many circumstances we can accept less accurate and more comparative results as being adequate for the task in hand. Very accurate measurements of interaction forces are possible in some simple systems. For instance, Evans et al (Chap. 9, this book) describe a very sensitive method that gives highly accurate measurements of the strength of a single adhesive bond. The methods detailed by Israelachvili et al (Chap. 3), Parker (Chap. 5) and Claesson (Chap. 2) in this book allow the determination of the whole force-distance relationship between layers of protein or lipids.

Cells are much more complex systems but they can be studied with simple and often rapid comparative methods such as those described by Bongrand and his collaborators in this book.

1.2.1 High Accuracy Quantitative Measurement

If we are interested in elucidating the mechanisms of adhesion in the finest detail we will need to make absolute measurements of the physical parameters of adhesion. Such measurements by themselves may be suggestive of particular mechanisms of adhesion but will normally be combined with precise chemical manipulation of the system to discover how, for example, changing the chemistry, e.g. receptor density or counter ion concentration, affects the system. In many cases there are multiple interpretations of the general effects of making a chemical change on an adhesion system. The quantitative measurement of adhesion allows you to exclude many of the alternative explanations.

In order to make these absolute measurements we will need to know the geometry of the contact. In this book the contributions by Curtis on interference reflection and other optical microscopies, by Gingell on the Total Internal Reflection Fluorescence Microscope and by Foa on the use of electron microscopy are designed to inform you on methods of obtaining information about the contact structure, even perhaps during the making or breaking of a contact. The force-distance relationships of the adhesive system will be measured in such systems so that the absolute magnitude of the energies as well can be calculated.

With the various techniques described by Evans, by Israelachvili, by Parker and by Claesson it is possible to obtain accurate information on intermolecular interactions including conformational effects and long-range interactions. So far these methods have only been applied to simple model systems which are not cellular or even good models of parts of cells. Whether the richness of events seen on the cellular level can be mimicked and correctly modelled and understood in model systems is yet to be demonstrated, but we are hopeful.

1.2.2 Relative Quantitative Methods

Such methods are those where you wish to compare the effects of some alteration of a system on its adhesive performance. Such investigations should, barring a few problems mentioned below, allow you to make statements that such and such a molecule is or is not involved in adhesion. The methods are relative and depend on the assumption that other features of the adhesion, for instance its geometry, have not changed when you alter the conditions in which you are interested.

1.2.3 Semi-Quantitative Methods

This type of measurement is only useful when it is being applied to some practical application. Questions such as 'Will polymer A allow cells to adhere more or less rapidly or extensively than polymer B ?' are questions whose answer may be valuable in a practical situation.

Or in some cases you are seeking the answer to some simple question such as 'What type of adhesive surface will allow cells to remain attached in a bioreactor when we stir the medium with a certain shear rate ?' No appreciable degree of understanding of the mechanisms of adhesion is necessarily required.

In these three types of methods, investigation can be looked at from the experimental point of view and lead to different degrees of control of the experimental set-up.

1.3 Types of Method

Methods used for measuring adhesion can be divided into two main ways.

One way in which we can divide the methods is to consider the methods used to perturb the systems. The two main methods used are (1) application of shear flow in an appropriate device and (2) application of measured and controlled forces through a microdevice.

The second way of classifying methods is to divide the topic into adhesion formation and adhesion breaking. In forming adhesions the extent to which adhesive forces perturb a collision between a cell and another or a cell and a non-living substratum are measured. In the latter the forces required to break the adhesion are estimated or measured. Adhesive forces also perturb this separation or attempted separation.

The following four sections will consider these methods of classification.

1.3.1 Perturbing Interactions by Flow Systems

The concept of perturbation may require some elucidation. Cells that meet in a shear flow will collide because a shear flow has different velocities at different positions. Generally these collisions will be two body interactions. If there is no interaction of attraction, i.e. adhesion between the cells, the shear flow will drag them apart after it brought them together. If there is some degree of attraction between the cells then the interaction will be perturbed and, if sufficiently perturbed, adhesions will form. This approach, a hydrodynamic one, in which the fluid forces acting to form or break the adhesion are estimated, has been a very powerful one and in this volume is instanced especially by the contributions by Goldsmith (Chap.10) and by Pierres et al. (Chap. 11), where the hydrodynamic forces acting on cells in suspension in a shear flow and on cells driven along a surface are analysed to show how they affect cell adhesion. It is worth recalling at this moment the methods introduced and analysed some years ago by Curtis (1969), Duszyk and Doroszewki (1986) and by Duszyk et al. (1986) for examining similar situations. Amblard describes the use of flow through an orifice to break cell adhesions. This again is a situation where the hydrodynamic regime is well characterised.

1.3.2 Perturbing Adhesions with Microdevice Systems

The other way of precisely perturbing the interaction is to drag the adhesion (cell to cell or cell to substratum) apart or force the adhesion together with microdevices or micropipettes. Evan Evans' paper describes elegant methods of doing this and Jacob Israelachvili reports on the investigation of the interactions of model adhesive surfaces where one surface is coated with biotin and the other with avidin, molecules known to interact strongly. John Parker and Per Claesson demonstrate how the surface force technique can be used to explore the effects on adhesion of conformation of adsorbed molecules, long range forces and the effects of divalent cations. Ragnar Erlandsson describes the atomic force microscope which is being used to characterise the lateral organisation of molecules on surfaces. The power of these methods lies in the detail of the information that can be obtained for simple model systems.

1.3.3 Formation Studies

Using adhesion formation systems is valuable if your interest is in how the adhesion first forms, and may be biologically very relevant in situations where cells form adhesions, egg-sperm, bacterial fouling, leukocyte or lymphocyte and endothelium. In one instance, the adhesion of platelets to the endothelium of the blood vessel described in interesting detail by Jan Sixma, the normal biological situation is close to an idealised measuring system. However, it should not be forgotten that biological events may change the nature of the adhesion soon after it is formed.

1.3.4. Break-Up Studies

In the second, the break-up of adhesions is examined. This immediately leads to the fact that normally we are investigating a very different biological situation from the formation situation. Usually the adhesion has been in existence for some appreciable time. Thus this method is useful if you are asking questions about the mechanism of adhesion in an established tissue. But even in the rare occasions when the adhesions are examined very rapidly after they have formed, the detachment (distraction, break-up) situation will not be the reverse of the formation situation because in one direction (formation) forces of attraction will be aiding adhesion and in the other tending to prevent separation. In addition there is the often overlooked and probably important drainage problem. This problem is simply that as two surfaces approach, the medium between them has to be drained out laterally and as they reseparate, medium has to be dragged into the opening gap. If the medium is appreciably viscous or contains charged species, the movement of the medium provides a force that apparently increases adhesion (if separation procedures are being used) or reduces adhesive forces if approach procedures (formation of an adhesion) are being studied. These important effects of hydrodynamic forces are also evident from surface force studies of the interaction between layers of mucin that cover many internal surfaces of the body.

1.4 Problems

There are many types of problem associated with the measurement of cell adhesion and these are referred to in many of the following contributions. They can be divided into :

- Problems of cell preparation see 1.4.1.-1.4.2 below.
- Possible damage or activation caused by the measuring method, see 1.4.1-1.4.3.
- Lack of knowledge of time-course of adhesion breaking or formation, see 1.4.4.
- Lack of knowledge of the geometry of the system, see 1.4.5
- Quantification of the drainage problem, see above and 1.4.6.

These sources of error can be handled as a series of specific problems.

1.4.1 Breakage of Membranes

The most extreme problem is one well known to technologists of the subject. Does an adhesion break at the level at which it forms? In the cellular context we can ask whether the plasmalemmae of two cells neatly peel apart or whether they rupture over appreciable areas. In the latter case the surfaces may look undamaged when you examine the separated cells but this can be simply because the plasmalemma flows over the damaged area, closes it, and reseals. In an extreme case it could be envisaged that the measurement is not of the adhesion, but of the strength of the plasmalemma to breakage.

Since mechanical shearing of cells can be used to porate cells to DNA for transfection experiments (Fechheimer et al. 1987), it is clear that extensive damage may occur without the cells dying.

1.4.2 Damage While Suspending Cells

If the adhesion formation situation is being investigated for tissue cells they will need to be dissociated (disaggregated) before the measurement can be made. Does the disaggregation method, which often involves the use of enzymes, damage the cells ? Luckily, cell types such as lymphocytes, leukocytes, and various tumour cells can be obtained in suspension form without the need to separate them from a solid tissue.

A half-way stage between the two types of investigation is provided by the collecting-lawn technique (Walther et al. 1973) where a monolayer of cells or even a thicker layer is provided to which a suspension of cells can attach. Naquet describes the use of such methods and Sixma places them in an in vivo context in the vascular system.

1.4.3 Activation of Cells

Cells, particularly leukocytes, may spend part of their lives in a non-adhesive state which can then be changed by some type of activation process to a highly adhesive state. The change may occur over a very short time period, possibly as short as seconds. Similar though reverse changes are possible whereby a cell suddenly loses adhesiveness.

In some cases treatment of the cells with various signal molecules or with antibodies or competitors for their receptors leads to changes in adhesion (e.g. Harlan and Liu 1992). One interpretation of such results is that the receptor and perhaps the signal molecule are involved directly in adhesion. It is interesting, for example, to note that many of the adhesion molecules have sequences, usually close to their transmembrane sequences, that are typical of molecules involved in signal reception at the surface. An alternative interpretation of their effects on adhesion is that these molecules have switched on (or off) mechanisms involved in setting a strongly adhesive state for the cells. If this alternative explanation is correct then it is possible when studying cell adhesion to misidentify indirect effects on adhesion as direct ones. Curtis et al. (1992 and in prep) showed that such mistakes have probably been made and suggests that many of the CAM molecules probably act in signalling pathways for the activation of adhesion.

There is some evidence that encounters of cells with particles, or certain types of substrate, may activate cell adhesion, and that exposure of the cells to serum switches off adhesion.

In any situation where large numbers of cells are used in suspension it should be borne in mind that the vessel in which they are contained has walls and that there may be appreciable interaction with the walls. Wang et al. (1993) have shown that slight mechanical perturbation of the plasmalemma may lead to the formation of focal contacts, see also Curtis and Clark (1990) and Forscher et al. (1992). In the two latter papers, interaction of the cells with micron width wires or with microbeads stimulated rearrangement of the cytoskeleton, which may well have altered cell adhesion. Thus bringing test particles into contact with cells may alter the adhesion the system was designed to measure.

1.4.4 Time Course of the Measurement

In many experiments the time for adhesion formation or breakage is not known, yet an accurate description of the time course may tell you a lot about the way the adhesion is made or is broken. If the change to a non-adhesive or to an adhesive state is slow, i.e. of the order of seconds, then all kinds of secondary effects may intervene (see immediately above). In the fluid flow situation at reasonable shear rates the time of the encounter can be measured or estimated and is normally in the order to 1 to 100 milliseconds which may be sufficiently short to escape the onset of secondary events. The paper by Goldsmith, see also Curtis and Lackie (1991), provides methods where this problem is probably obviated because the lifetime of the interaction is too short for the cell to react.

1.4.5 Cell Shape

It is usual to treat cells as being of simple shape so that the mechanics of separation or approach may be uncomplicated. In fact, cells are rarely of simple shape so that our modelling systems may well give incorrect results. Use of techniques such as interference reflection microscopy provides a method for looking at the geometry of the contact zone during adhesion formation or breakage. The modifications to the hydrodynamic equations that then become necessary may give computational difficulty.

1.4.6 The Drainage Problem

This has been mentioned above. It is an important feature rarely mentioned by investigators in the field. For example, if the viscosity of the suspending medium is changed, then the hydrodynamics of the situation are altered. If systems are compared with a different and higher viscosity of suspending medium, such as might be caused by adding a protein   solution to a system,   then   cell   separation   measurements   will   suggest   higher adhesiveness and attachment experiments lower adhesion when in fact the cell adhesiveness may not have changed. Since the medium is flowing in or out of a narrow zone with charged walls (the plasmalemma), electroviscous effects may also occur.

1.5 Comment on Less Precise Methods

Though this book was planned to include only the more precise methods, there are many methods in use where the measurement is decidedly imprecise. These include methods whereby the diameter of aggregates formed is used as a measure of adhesion, or the number of cells captured by a surface in a certain time period or changes in the optical density of cell suspensions are measured to estimate adhesiveness. The reasons for the defects in such methods lie amongst the following:

i. Overlong time course of measurement.
ii. Failure to control effects of medium viscosity, see drainage problem above.
iii.Failure to take account of changes or differences in cell size or shape.
iv.Failure to   carry out the measurement in a defined and known hydrodynamic situation.
v. Failure to take account of the effects of cell population density on aggregation kinetics.

1.6 Prospects

Until recently, studies on cell adhesion, despite an apparent air of success, were unknowingly marred by difficulties in separating adhesion effects from related effects. Methods for solving this problem are now available and are discussed in this book. In many ways the aim of the book is to look forward and to inspire the use of better methods rather than to report on what has happened in the past. We hope that you, the reader, will be inspired. One of the main goals of biologists working in this area is to be able to account for the strength of adhesions in molecular terms. This is beginning to be attainable.

1.7 Detailed Description of the Organisation of This Book

An unusual feature is the incorporation of physical methods that are rarely used by biologists at the present time. The rationale of this choice was that these methods may give new information on biological molecules or even actual or model cells within the next few years. Therefore, it is probably now a suitable moment to try and apply them to biological material.

The Surface Force Apparatus provides a unique way to measure at the same time the interaction force (with a sensitivity of the order of 10 micronewton) and the distance (with about 0.1 nanometer accuracy) between smooth surfaces. The general background is presented by Per Claesson, and applications to specific bonding between biological molecules are described by Jacob Israelachvili (who pioneered the SFA for more than 20 years).

Two basic limitations of this approach are that measurements are fairly slow and only average values representing multiple molecular interactions can be studied. These limitations might in principle be partially overcome by adapting scanning tunnelling microscopy or atomic force microscopy to obtain sensitivities better than 100 piconewton. Although these methods still require some improvement before they can be used regularly and reliably by biologists, it is clear that biologists will have to start using them to develop that improvement. Erdlandsson and Olsson present a general review of the present methods and difficulty of interpretation. John Parker describes a very interesting adaptation that may allow much more rapid measurements, thus meeting the need to understand the behaviour of cell surface molecules under dynamical conditions.

The following four chapters describe methods for determination of the strength of cell-cell and cell-substratum adhesion. Indeed, there is an obvious need for scientists to use a clear definition of the phenomena they study, and a cell may be defined as bound to another cell only by showing that these cells cannot be separated by a force of intensity F applied during a given time t. Anne Marie Benoliel et al. describe a very affordable methodology allowing semi-quantitative determination of the strength of adhesion based on the use of hydrodynamic flow. Cells can be subjected to a wide range of separating forces (between 1 and 100,000 piconewton) for a short time (typically less than one second). The basic limitation is that the force field is not uniform and different cells may experience forces of markedly different intensities. Francois Amblard describes a very powerful adaptation of flow cytometry allowing this difficulty to be overcome. Cell-cell conjugates are subjected to a well-defined calibrated force during a short time and immediately counted. It is thus possible to assay numerous conjugates during a short time. This method requires a slight modification of the cytometer, which may be difficult to achieve in a biological laboratory. The limitation of aforementioned methods is that the duration of the mechanical separation is difficult to control. Philipe Naquet describes a different approach based on centrifugation of bound cells. This is an adaptation of a procedure first described by McClay et al. (1981) that allows quantitative study of cell-surface adhesion strength. The range of force intensities is very wide (from 0.1 piconewton to several nanonewton), and the duration can be varied at will, although it is difficult to make it shorter than several seconds. This may be easily performed with standard biological equipment. Finally, Evan Evans presents an ultrasensitive method allowing careful control of the rupture of a single molecular attachment to a cell surface, based on micromanipulation techniques. The range of force measurable is very wide (between 0.1 piconewton and 1 nanonewton) and the kinetics of separation can be determined with high accuracy. However, this approach requires sophisticated apparatus, and it is quite slow. It is therefore not suitable to study large numbers of cells under varying experimental situations.

In the next chapter, Harry Goldsmith presents the sophisticated travelling microtube apparatus he developed for about 15 years. This allows very accurate monitoring of bond formation and rupture between individual spherical particles coated with binding molecules. Single molecular attachments can be studied (indeed, this approach allowed the first experimental estimation of the strength of single molecular bonds by Tha et al. 1986). However, this requires custom made apparatus, and it may be difficult to apply to nonspherical particles. The following chapters describe the use of parallel-plate flow chambers to study cell-substratum adhesion. Anne Pierres and colleagues present a computer-assisted approach allowing accurate analysis of the trajectories of cells driven along adhesive surfaces by a laminar shear flow. The presentation of some methodological drawbacks is of general interest. It is hoped that this may be useful to determine basic kinetic parameters of bond formation and dissociation. Then Jan Sixma describes a flow chamber he developed for 10 years. The biological model he presents (platelet adhesion) is ideally suited to demonstrate the importance of studying dynamical bond formation, since the behaviour of a given adhesion molecule may be very different under static and dynamic conditions.

Quantitative determination of the rate of bond formation and dissociation is certainly insufficient to provide a molecular understanding of the mechanisms of cell adhesion. Indeed, cells are complex objects with quite irregular shape and a surface coated with numerous molecular species. The last three chapters are devoted to complementary methods that may provide useful information on cell structure and function in relation to adhesion. Adam Curtis describes the potential of interference reflection microscopy. This methodology, which he pioneered 20 years ago, allows semi-quantitative determination of cell-substratum distance with a fairly standard microscope. In the next chapter, David Gingell describes a complementary approach based on the use of evanescent waves, that may also be used to study cell-substratum distance. The last chapter by Colette Foa and colleagues is devoted to various applications of electron microscopy. Indeed, whereas the preparation of samples for electron microscopy is fairly tedious and may lead to artefacts, it is certainly the most powerful way of obtaining structural information on binding areas, since intercellular gaps can be measured with better than 10 nanometer accuracy, together with local density of adhesion molecules (provided suitable labelling is performed) or atomic species (using X-ray microanalysis or electron energy loss spectroscopy). Specific applications are presented. It is also shown how additional information can be obtained with easily affordable computing devices, provided custom software is developed.

Acknowledgement The origin of this book stems in an INSERM workshop held at Le Vésinet, June 2-3, 1993. We thank INSERM for their assistance in making the meeting possible.

References

Curtis ASG (1969) The measurement of cell adhesiveness by an absolute method. J.Embryol Exp Morph 22:305-325

Curtis ASG, Clark P (1990) The effects of topographic and mechanical properties of materials on cell behavior. CRC Revs Biocompatibility 5: 343-362

Curtis ASG, Lackie JM (eds) (1991) Measuring Cell Adhesion. John Wiley & Sons, Chichester

Curtis ASG, McGrath M, Gasmi L (1992) Localised application of an activating signal to a cell: experimental use of fibronectin bound to beads and the implications for mechanisms of adhesion. J Cell Sci 101: 427-436

Duszyk M, Doroszewski J (1986) Poiseuille flow method for measuring cell-to-cell adhesion. Cell Biophysics 8: 119-130

Duszyk, M, Kawalec M, Doroszewski J (1986) Specific cell-to-cell adhesion under flow conditions. Cell Biophysics 8:131-139

Edelman GM, Thiery J-P(eds) (1985) The cell in contact.   Adhesions and junctions as morphogenetic determinants. Wiley, New York.

Ellwood DC, Melling J, Rutter P. (1978) Adhesions of micro-organisms to surfaces. Academic Press, London

Fechheimer M, Boylan JF, Parker S., Sisken JE, Patel JL, Zimmer SG (1987) Transfection of mammalian cells with plasmid DNA by scrape loading and sonication loading. Proc Natl Acad Sci USA 84 : 8463-8467

Forscher P, Lin CH, Thompson C (1992) Novel form of growth cone motility involving site-directed actin filament assembly. Nature 357: 515-518

Harlan. JM, Liu DY (eds) (1992) Adhesion. Its role in inflammatory disease. W.H.Freeman, New York

McClay DR., Wessel GM, Marchase RB (1981) Intercellular recognition : quantification of initial binding events. Proc Natl Acad Sci USA 78:4975-4979

Tha SP, Shuster J, Goldsmith HL (1986) Interaction forces between red cells agglutinated by antibody. II. Measurement of hydrodynamic forces of breakup. Biophys J 50:1117-1126

Walther BT, Ohman R, Roseman S (1973) A quantitative assay for intercellular adhesion. Proc Natl Acad Sci USA 70:1569-1573

Wang N, Butler JP, Ingber DE (1993) Mechanotransduction across the cell surface and through the cytoskeleton. Science 260 : 1124-1126

2 The Surface Force Technique: Interactions Between Surfactant Bilayers and Between Protein Layers

P. M. Claesson

2.1 Introduction

The forces operating between colloidal particles are of utmost importance for the stability of dispersions and emulsions. Such forces also determine the adhesion between macroscopic surfaces, the adsorption of colloids onto surfaces and they are important in the initial stage of the cell adhesion process. In applied surface chemistry much effort is going into modifying the forces acting between particles in order to give the often complex technical colloidal systems the preferred properties. This has inspired a great deal of fundamental research concerned with determining which types of forces are operating and how the strength and magnitude of these forces can be varied by addition of additives to the solution or by surface modification techniques.

This chapter will start with a brief recapitulation about the types of forces that are important in colloidal systems and some basic properties of these forces. Next, one important method for studying these forces, the interferometric surface force technique, will be described. Finally, the type of information that can be obtained with this technique will be discussed using examples from our studies of surfactant bilayers and of adsorbed protein layers.

2.2 Colloidal Forces

2.2.1 Van Der Waals Forces

Van der Waals forces originate from the movement of negatively charged electrons around the positively charged atomic nucleus. For condensed materials (liquids or solids) this electron movement gives rise to a fluctuating electromagnetic field that extends beyond the surface of the material. Thus when, e.g. two particles or a cell and a surface are getting close together the fluctuating fields associated with them will interact with each other. The free energy of interaction per unit area (W_{vdw}) between two flat surfaces a distance D apart is given by:

$$W_{vdw} = - \frac{A}{12\pi D^2} \tag{1}$$

where A is the so called nonretarded Hamaker constant that depends on the dielectric properties of the two interacting particles and the intervening medium. When these properties are known one can calculate the Hamaker constant. An approximate equation for two particles (subscript 1 and 3) interacting across a salt-free medium (subscript 2) is:

$$A = \frac{3kT}{4}\left(\frac{\varepsilon_1 - \varepsilon_2}{\varepsilon_1 + \varepsilon_2}\right)\left(\frac{\varepsilon_3 - \varepsilon_2}{\varepsilon_3 + \varepsilon_2}\right)$$
$$+ \frac{3h\nu}{8\sqrt{2}}\frac{\left(n_1^2 - n_2^2\right)\left(n_3^2 - n_2^2\right)}{\sqrt{n_1^2 + n_2^2}\sqrt{n_3^2 + n_2^2}\left[\sqrt{n_1^2 + n_2^2} + \sqrt{n_3^2 + n_2^2}\right]} \tag{2}$$

where

- k = Boltzmann's constant
- T = the absolute temperature
- ν = the main absorption frequency in the UV region, often about 3×10^{15} Hz
- h = Planck's constant
- ε = the static dielectric constant
- n = the refractive index in visible light

We note that a positive Hamaker constant results in a negative (attractive) energy of interaction. Here we have neglected retardation effects, which are due to the finite speed of the propagation of electromagnetic waves. Such effects will lower the van der Waals force at large separations compared to predictions based on eq.1. When salt is added to the intervening medium it results in a decrease of the first term in eq.2 whereas the second term is unaffected (Davies and Ninham 1972). For further discussions of the approximations made in order to obtain eq. 2 the reader is referred to Israelachvili (1991).

An inspection of eq. 2 allows several conclusions to be drawn:

i) The van der Waals interaction between two particles of the same kind is always attractive whereas it can be repulsive between different types of particles.

ii) The magnitude of the van der Waals attraction increases with the difference in dielectric properties between the medium and the particles.

iii) For particles or cells interacting with each other or a foreign surface in water the van der Waals interaction will be less important if the cell and/or the foreign surface is coated with a layer that contains a large quantity of water (e.g. polysaccharide layers).

2.2.2 Electrostatic Double-Layer Forces

Electrostatic double-layer forces are always present between charged surfaces in electrolyte solutions. Counterions to the surface (ions with opposite charge to that of the surface) will be attracted to the surface and co-ions repelled. Hence, outside the surface, in the so-called diffuse layer, the concentration of ions will be different from that in bulk solution, and the charge in the diffuse layer balances the surface charge. An electrostatic double-layer interaction arises when two charged surfaces are so close together that their diffuse layers overlap. When the electrostatic surface potential is small the free energy of interaction per unit area between flat surfaces (W_{dl}) is given by (Parsegian and Gingell 1972):

$$W_{dl} = \frac{\kappa^{-1}}{\varepsilon_0 \varepsilon} \frac{\left(\sigma_1^2 + \sigma_2^2\right)e^{-\kappa D} + 2\sigma_1\sigma_2}{\left(e^{\kappa D} - e^{-\kappa D}\right)} \qquad (3)$$

which for large distances reduces to:

$$W_{dl} = \frac{2\kappa^{-1}\sigma_1\sigma_2}{\varepsilon_0 \varepsilon} e^{-\kappa D} \qquad (4)$$

where
- ε_0 = the permittivity of vacuum
- ε = the static dielectric constant of the medium
- σ = the surface charge density
- κ^{-1} = the Debye screening length given by:

$$- \quad \kappa^{-1} = \sqrt{\frac{\varepsilon_0 \varepsilon kT}{1000 N_A e^2 \sum_i c_i z_i^2}} \tag{5}$$

where
- e = the electron charge
- N_A = Avogadro's number
- c_i = the concentration of ion i expressed in mol/dm^3
- z_i = the valency of ion i

Provided that the surface charge is independent of surface separation one can draw the following conclusions from eqs. 3-5.

i. At large separations the double-layer interaction decays exponentially with surface separation and the decay length equals the Debye length.
ii. The Debye length and consequently the range of the double-layer force decreases with increasing salt concentration and the valency of the ions present.
iii. At large separations, surfaces having the same sign on their charge repel each other whereas surfaces having opposite sign on their charge attract each other.
iv. Surfaces having opposite sign on their charge repel each other at sufficiently small separations provided that the magnitude of the surface charges are not the same.

A further complication is that the surface charge density for both surfaces may vary with the surface separation which may result in a very complex distance dependence of the double-layer force (Chan et al.1976). The famous *DLVO theory* for colloidal stability (Verwey and Overbeek 1948) takes into account double-layer forces and van der Waals forces.

2.2.3 Hydration / Steric-Protrusion Forces

Hydration and steric-protrusion forces are repulsive forces that have been found to be present at rather short separations between hydrophilic surfaces such as surfactant headgroups. At least two molecular reasons for these forces have been identified. First, when two polar surfaces are brought close together, the polar groups will be partly dehydrated, which gives rise to a repulsive force (Parsegian et al. 1979). Secondly, as two surfaces are brought close together, the molecules at the interface will have a decreased mobility perpendicular to the surface, which decreases the entropy of the system and therefore also gives rise to a steric type of repulsion (Israelachvili and Wennerström 1992). Empirically it has been found that the hydration/steric repulsion between surfactant and lipid headgroups decays roughly exponentially with distance.

$$W_{hyd} = W_{hyd}^0 e^{-D/\lambda} \tag{6}$$

where λ is the decay length of the force, typically 0.2-0.3 nm.

2.2.4 Hydrophobic Forces

Hydrophobic attraction forces between macroscopic hydrophobic surfaces have been found to be very long-ranged. Such forces have been experimentally observed for a range of different hydrophobic surfaces prepared by adsorption from solution (Israelachvili and Pashley 1984; Pashley et al. 1985), Langmuir-Blodgett deposition (Claesson et al. 1986a ; Claesson and Christenson 1988) or chemical silanation reactions (Rabinovich and Derjaguin 1988 ; Parker et al. 1989a). At short separations the measured force decays roughly exponentially with a decay length of 1-3 nm.

$$W_{hph} = W_{hph}^0 e^{-D/\lambda} \tag{7}$$

At large separations the attraction decays less rapidly. That the contact attraction between hydrophobic surface in water should be strongly attractive is well understood. However, there is no agreement in the scientific community about the mechanism that can give rise to such a *long-range* attractive force between hydrophobic surfaces. In fact, this point remains as one of the great challenges in colloid and surface science today. Relevant is that experimentally it has been observed that the adsorption of one layer of surfactants with the polar headgroups towards the solution on top of a hydrophobic surface completely removes the long-range attraction (Claesson et al. 1986b ; Herder 1991). Further, studies of surfaces exposing a variable ratio of CH_3 and $COOH$ groups towards the solution show that the long-range attraction, in excess of the van der Waals force, is removed when the ratio of CH_3 to $COOH$ groups is close to one (Berg and Claesson 1989). A further conclusion is that the strength of the attraction is not simply proportional to the exposed hydrocarbon area but is much more dramatically reduced when polar groups are introduced onto hydrophobic surfaces. When considering the adsorption of cells onto solid surfaces it is clear that the cell does not expose very large hydrophobic regions and one should not expect the very long-range attractive forces observed between homogeneously hydrophobic surfaces to be present. However, at short separations hydrophobic forces are likely to be important provided that the solid surface is partly hydrophobic. So far only a few studies concerned with the forces acting between one polar and one nonpolar surface have been reported (Claesson et al. 1987 ; Tsao et al. 1993 ; Parker and Claesson 1994). In all cases the short-range force was attractive.

2.2.5 Polymer Induced Forces

The presence of polymers on surfaces gives rise to additional forces, which can be repulsive or attractive. Under conditions when the polymer is firmly anchored to the surface and the surface coverage is large, a *steric repulsion* is expected. As the surfaces are brought together, the segment density between them increases, which results in an increased number of segment-segment contacts and a loss of conformational entropy of the polymer chains. The conformational entropy loss always results in a repulsive force contribution that dominates at small separations. The increased number of segment-segment contacts may give rise to an attractive or a repulsive force contribution. This is often discussed in terms of the chi-parameter (χ-parameter) or in terms of solvent quality. Under sufficiently poor solvent conditions ($\chi > 1/2$), when the segment-segment interaction is sufficiently favourable compared to the segment-solvent interaction, the long-range interaction can be attractive otherwise it is repulsive.

When the surfaces are not completely coated with polymers, a *bridging attraction* between the surfaces may take place. This occurs when polymers adsorbed on one surface also have segments that experience an attractive force from the other surface. For uncharged polymers this most often means that the polymer should bind to both surfaces. For polyelectrolytes, however, that may experience a long-range electrostatic attraction to the surface, it is sufficient for the polyelectrolyte to be close to, but not necessarily bind to, both surfaces to cause attraction (Åkeson et al. 1989). An extreme case of bridging attraction is when one polymer coated surface is approached towards a non-polymer coated surface (Dahlgren et al. 1993). In this case the long-range force will be attractive provided that the polymer segments have a sufficient affinity to the surface. This affinity may arise from hydrophobic interactions between polymer segments and the surface, electrostatic interactions or the types of specific interactions discussed below. The short-range force will again be repulsive due to the loss in polymer chain entropy. A typical example of this type of interaction is given in Fig. 1 that shows the forces acting between one negatively charged mica surface and one mica surface having a small positive charge due to adsorption of a cationic polymer.

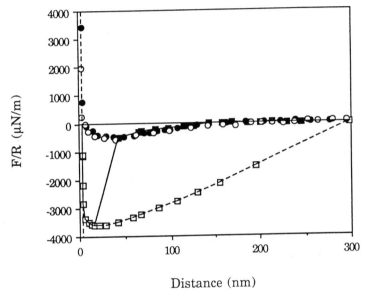

Fig. 1. The force curve between one bare mica surface and one mica surface coated with a cationic polyelectrolyte across a 0.1 mM KBr solution. ■, first approach ; □, first separation ; ●, second approach ; ○, third approach.

2.2.6 Specific Forces

Finally there is a range of different short-range forces that occur between pairs, or triplets, of molecules. Most important among these may be hydrogen bonds, acid/base interactions and ionic bridges, e.g. Ca^{2+} ion mediated bridges between two carboxylic groups. These types of forces are important between surfaces in contact. The range of these forces is too short to allow any determination of the distance dependence using the surface force technique. Nevertheless, their effect on adhesion and adsorption is easily detected.

2.3 Surface Force Technique

With the surface force technique developed by Israelachvili et al. (Israelachvili and Adams 1978 ; Parker et al. 1989b) the total force acting between two macroscopic (A $\approx$ 1cm^2) molecularly smooth surfaces in a crossed cylinder configuration is measured as a function of surface separation. This choice of geometry is made due to its experimental convenience. There will be only one contact position, which easily can be changed by moving the surfaces, and there is no problem with aligning the surfaces. One of the surfaces is mounted on a piezo electric tube that is employed to change the surface separation. The other surface is mounted on the force measuring spring.

Muscovite mica, a layered aluminosilicate mineral, is the preferred substrate in these measurements due to the ease with which large molecularly smooth thin sheets can be obtained. The surface chemical properties of these surfaces can be modified in a number of ways including adsorption from solution, Langmuir-Blodgett deposition, and plasma treatment (in some cases followed by reactions with chlorosilanes). The mica sheets are silvered on their back-side and glued onto half-cylindrical silica discs (silvered side down), normally using an epoxy glue. When white light is introduced perpendicular to the surfaces, an optical cavity is formed between the silvered backsides of the mica surfaces. From the wavelengths of the standing waves produced the surface separation can be measured to within 1-2 Å. The deflection of the force measuring spring can also be determined interferometrically, and the force is calculated from Hooke's law.

2.3.1 Typical Experimental Procedure

The procedure used when studying the interactions between protein layers adsorbed to solid surfaces is briefly described below. The experimental results are sensitive to contaminants, particularly particles adsorbed to the surfaces, and therefore it is advisable to always make some control experiments before the actual research is started.

Once the two surfaces are mounted in the surface force apparatus they are brought into contact in dry air. When the surfaces are uncontaminated the long-range force is due to the attractive van der Waals force and the adhesion between the surfaces is high. In the next step, the measuring chamber is filled with pure water or a weak aqueous electrolyte solution. Under these conditions it is well established that the long-range force is dominated by a repulsive electrostatic double-layer force and the short-range interaction by a van der Waals attraction, and that the measured forces are consistent with theoretical predictions based on the DLVO theory. Hence, it is easy to establish that the measured forces at this stage are as expected in an uncontaminated system. The adhesive contact between the surfaces in water (i.e. the wavelengths of the standing waves with the surfaces in contact) defines the zero surface separation. When everything appears as normal, the salt concentration and pH are changed to the values preferred under the experiment. The forces are measured a few times and the results are checked for reproducibility and compared with previous results (if available). It is only at this stage that the protein is introduced into the measuring chamber and the true research can be started.

2.3.2 What Is Really Measured?

Before discussing some experimental results, I would like to point out a few things that may not be obvious for scientists with no previous experience with this technique.

First, the force is not measured between individual molecules but rather all molecules on the surfaces over an area of about 100 μm^2 contribute to the force. Hence, the measured total force reflects an average of all these interactions that will depend on the orientation of the molecules. When the driving force for adsorption of one part of the molecule is very different from that of the other part, e.g. surfactants on a hydrophobic surface, one orientation of the molecules will dominate. In other cases, like for proteins on surfaces, one can expect that different orientations on the surface are possible. If the protein is strongly asymmetric it means that the long-range part of the force curve could be dominated by a few proteins adsorbed end-on even though most molecules are adsorbed side-on. This has for example been observed to be the case for human serum albumin adsorbed on mica from a 0.15 M NaCl solution (Fig. 2).

The force (F_c) is measured between crossed cylinders with a geometric mean radius R, which is related to the free energy of interaction between flat surfaces per unit area (W) (Derjaguin 1934 ; Israelachvili 1991):

$$\frac{F_c}{R} = 2\pi W = 2\pi \sum_i W_i \qquad (8)$$

where W_i represents the different types of interactions as discussed above.

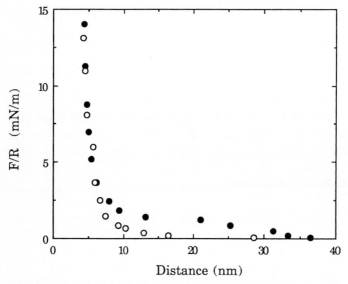

Fig. 2. The force curve between mica surfaces coated with human serum albumin (HSA) across a 0.15 M NaCl solution containing 0.01 mg/ml HSA. ● , first approach ; ○ , second approach.

This relation is valid provided that R (about 2 cm) is much larger than the range of the forces between the surfaces (typically 10^{-5} cm or less). A requirement fulfilled in this experimental set-up. The local radius can be measured from the shape of the standing wave pattern. Experimentally it is found that under the influence of strong forces, particularly when the forces vary rapidly with surface separation, the local radius will change as the surfaces are pushed firmly together. This is due to the flattening of the glue supporting the mica surfaces. (Note that the measured surface separation is not influenced by the deformation of the glue since it is located outside the optical cavity). When the shape of the surfaces is deformed, the Derjaguin approximation is no longer valid. This problem, that complicates the interpretation of the measured forces, has recently been treated theoretically (Parker and Attard 1992).

An increased understanding of the results can often be obtained by interpreting the measured total force in terms of various force contributions. However, it should be stressed that this procedure in itself is an approximation since all forces are interrelated and not strictly independent. Nevertheless, it is often a useful approximation. The most common procedure is to calculate the electrostatic double-layer force and the van der Waals force for the system under investigation. The discrepancy between measured and calculated forces is interpreted in terms of other force contributions.

2.4 Examples and Data Interpretation

2.4.1 Surfactant Bilayers

Surfactant bilayers can be obtained on mica by adsorption from solution (soluble surfactants) or by Langmuir-Blodgett deposition ("insoluble" surfactants). Here, we will consider the forces operating between

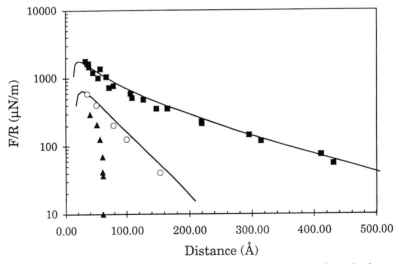

Fig. 3. The forces between dihexadecylphosphate layers deposited on hydrophobized mica. The forces were measured in aqueous $CaCl_2$ solutions. ■ , 10^{-4} M ; ○ , 10^{-3} M ; ▲ , 10^{-2} M. Lines are calculated DLVO force curves.

dihexadecylphosphate (DHP) layers deposited on hydrophobized mica surfaces. In these layers the polar phosphate groups are exposed towards the aqueous solution. The long-range forces operating between the DHP layers at pH about 6 are shown as a function of the concentration of CaCl$_2$ in Fig. 3.

The most long-range part is caused by an electrostatic double-layer force originating from the negatively charged DHP. As expected, the range of the double-layer force decreases as the CaCl$_2$ concentration is increased, and the decay-length of the force is consistent with theoretical predictions. It is also observed that the apparent surface charge density, obtained from a best fit of the DLVO theory to experiments, decreases with increasing CaCl$_2$ concentration, indicating that calcium ions adsorb to the DHP monolayer.

Figure 4 shows the adhesion force measured between the DHP layers under various solution conditions (pH about 6).

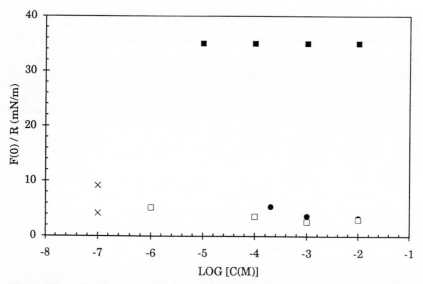

Fig. 4. The adhesion force between dihexadecylphosphate layers deposited on hydrophobized mica in the presence of various cations. CaCl$_2$ solutions ■ , MgCl$_2$ solutions ; □ , NaCl solutions ; ● , pure water, x.

The results are very different depending on the cation present in solution. Calcium ions increase the adhesion significantly, whereas sodium ions and magnesium ions decrease the adhesion between the DHP layers. It appears that phosphate - Ca^{2+} - phosphate bridges are formed. It is no surprise that similar bridges are not formed in the presence of Na$^+$, but the difference between Ca^{2+} and Mg^{2+} is remarkable. The reason is most likely that Mg^{2+} is much more hydrated and therefore has less to gain by binding to the phosphate group than Ca^{2+}. Calcium ions also reduce the headgroup repulsion *within* DHP layers, whereas magnesium ions increase the repulsion, as seen from surface pressure area isotherms.

The presence of divalent ions also affects the adhesion between surfaces exposing carboxylic acid groups. For example, the adhesion at pH 7 between deposited layers of arachidic acid, exposing the carboxylic acid groups, was found to be 10-20 mN/m in water. Upon addition of $CdCl_2$ to a concentration of 3×10^{-4} M, the adhesion force increased by a factor of about 20 (Claesson and Berg 1989). Another example of the adhesion promoting effect of (some) divalent ions is found when studying poly(acrylic acid) adsorption on mica. In the absence of Ca^{2+} ions no adsorption occurs. However, PAA adsorbs readily when $CaCl_2$ is added to a concentration of 3×10^{-3} M (Berg et al. in press). Also, it is observed that the adhesion between the adsorbed PAA layers increases with increasing pH, presumably because more bridges are formed.

Many cell surfaces expose polysaccharides which are negatively charged due to the presence of, among other things, carboxylic acid groups. It seems plausible that in the presence of calcium ions these groups may promote the adhesion to negatively charged surfaces.

2.4.2 Difficulties Associated with Studies of Deposited Bilayers Using the Surface Force Technique

The deposited bilayers are non-equilibrium structures, which means that it may be expected that they are rather unstable. For instance, they may slowly dissolve into the aqueous phase. It has been shown that this process can be slowed down or stopped by saturating the water with the surfactant prior to filling the experimental chamber (Marra and Israelachvili 1985). This can have some disadvantages if the solution condition is changed in such a way that the headgroup interaction becomes strongly attractive (for instance by adding divalent ions) in which case multilayers may start to build up on the surfaces. For monoolein it has been observed that the deposited layer with time is converted into another structure (Pezron et al. 1991), which presumably is related to that found in the cubic phase (which is the phase in equilibrium with pure water).

Under conditions where the deposited layers adhere strongly to each other they may also be destroyed when the adhesive contact is broken (Claesson and Berg 1989). The reason for this is that the binding energy between the deposited layer and the surface is of the same order of magnitude as the adhesion between the layers that are brought into contact. This breakdown is clearly seen when the forces are measured a second time.

2.4.3 Protein Coated Surfaces

As mentioned above, it is not expected that proteins adsorb on surfaces in equally well defined layers as surfactants. Rather, one would expect a range of different orientations and a less homogeneous surface coverage. Despite this, the surface force technique can provide valuable information about orientation, multilayer formation and large conformational changes for proteins on surfaces.

Human serum albumin (HSA) is a globular but rather flexible protein with a molecular weight of 66000 g/mol and an overall dimension of 14x4x4 nm. It consists of three rather compact units held together by

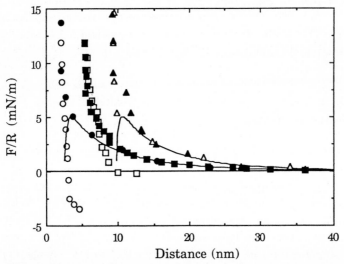

Fig. 5. The force curve between mica surfaces coated with human serum albumin (HSA) across a 1 mM NaCl solution at pH 5.5 containing 0.001 mg/ml HSA (●,○); 0.01 mg/ml HSA (■,□); 1 mg/ml HSA (▲,△). *Filled symbols* represent forces measured on approach, *unfilled symbols* forces measured on separation. *Solid lines* are calculated DLVO forces.

short flexible regions. The adsorption of HSA at a low ionic strength (10^{-3} M NaCl) at pH 5.5, where it is weakly negatively charged, onto negatively charged mica and the resulting surface forces have been investigated (Blomberg et al. 1991). First, we note that the protein do adsorb onto mica despite the fact that it has the same net sign on its charge as the surface.

The forces measured depend strongly on the HSA concentration in bulk solution as illustrated in Fig. 5.

In all cases the long-range force is dominated by a repulsive double-layer force originating from the charges on the mica surface and on the HSA. When the HSA concentration is 0.001 mg/ml, a hard wall repulsion is encountered at a separation of about 2 nm. This distance is less than the smallest cross-section of HSA. Hence, it is clear that when the number of molecules adsorbed on the surface is small the conformation of the protein can easily be changed by an external compressive force, indicating a rather limited structural stability. On separation, an attractive minimum is observed at a distance of about 4-5 nm. This corresponds to the expected thickness of one monolayer of HSA adsorbed side-on. As the concentration is increased further to 0.01 mg/ml the range of the non-DLVO force increases to about 7-8 nm, close to the expected thickness for one side-on monolayer adsorbed on each surface.

When a high compressive force is applied the thickness decreases to about 5-6 nm. No, or a weak, adhesion force is observed between the layers.

When the HSA concentration is even larger, 1 mg/ml, the range of the non-DLVO repulsion is 12-15 nm. This is considerably larger than expected for a side-on monolayer on each surface, demonstrating that some molecules are adsorbed at an angle to the surface. Under the action of a high compressive force the layer thickness is reduced to about 8-9 nm. No adhesion between the surfaces is observed under these conditions.

It was also observed that the protein layer was not desorbed when the protein solution was replaced with a protein-free salt solution. This made it possible to replace one of the protein coated surfaces with one bare mica surface and measure the surface forces. A comparison between the forces measured before and after replacing one of the protein coated surfaces with bare mica is shown in Fig. 6. As expected, replacing one of the protein coated surfaces with bare mica results in a halving of the contact separation. More interesting is that the adhesion force between the mica surface and the protein coated surface is high, much higher than seen between two HSA coated surfaces. This suggests that the adhesion force observed between albumin coated surfaces with a low protein coverage is due to bridging of proteins between the two surfaces rather than to attractive interactions between partly denatured proteins.

Lysozyme is a small compact protein with dimensions approximately 4.5x3x3 nm. It is positively charged at pH 6, the pH used in this study.

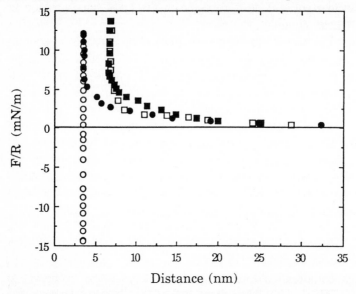

Fig. 6. Force curves across 1 mM NaCl, pH 5.6. The forces were measured between two mica surfaces precoated with HSA (adsorption from a 0.01 mg/ml HSA solution) (■, □), and between one bare mica surface and one surface coated with (HSA) (●, ○). *Filled symbols* represent forces measured on approach, *unfilled symbols* forces measured on separation.

The positive charges on the molecule are located in such a way that it is difficult to use electrostatic considerations to predict the orientation of lysozyme on negatively charged surfaces. When the lysozyme concentration is 0.2 mg/ml the adsorption on mica is such that the mica lattice charge is essentially neutralised (Tilton et al., 1993). Hence, no or a very weak double-layer force exists between the lysozyme-coated surfaces. The forces acting between the surfaces are shown in Fig. 7.

Already after a short adsorption time the surface charge is neutralised and no force is experienced until the surfaces are about 14 nm apart. From this point an attractive force pulls the protein-coated surfaces together. A hard-wall repulsion is located at a separation of 6-7 nm. This corresponds to one side-on monolayer on each surface. When the surfaces are compressed firmly together no change in surface separation is observed. Hence, no indication of a surface or pressure induced denaturation is observed for lysozyme, which demonstrates that the conformational stability is considerably larger than that for HSA. For longer adsorption times the adsorbed layer thickness increases to about 9 nm, corresponding to an end-on monolayer on each surface. Also, a weak non-electrostatic repulsion is observed at separations between 12 and 10 nm. These findings indicate that with time a significant fraction of the adsorbed molecules are in an end-on orientation (the result does not show that all, or even most of, the molecules are oriented end-on). It

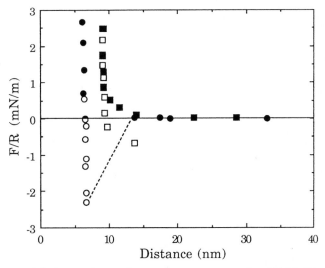

Fig. 7. Forces measured between two mica surfaces across a 1 mM NaCl solution at pH 6 also containing 0.2 mg/ml lysozyme. Equilibrium forces are represented with (■, □), and forces measured after a short adsorption time with (●, ○). *Filled symbols* correspond to forces measured on approach and *unfilled symbols* to forces measured on separation.

also appears that a second layer of lysozyme is weakly associated with the firmly bound one. It is the work needed to remove this layer from the contact zone that gives rise to the repulsion observed between 12 and 10 nm. A similar, but more dramatic, build-up of the protein layer has been observed for insulin, a small protein that readily associates into dimers, tetramers and hexamers in solution (Claesson et al. 1989b).

A completely different interaction pattern is observed for surfaces coated with mucin (Malmsten et al. 1992), a large strongly hydrated glycoprotein which covers many internal surfaces in the body. In this case steric and hydrodynamic forces are most important.

2.4.4 Difficulties Associated with Studies of Protein Layers Using the Surface Force Technique.

Most of the difficulties associated with the studies of adsorbed proteins are concerned with the interpretation of the measured forces. The dielectric properties of the adsorbed layer will not be accurately known which makes it impossible to exactly determine the van der Waals force. Further, due to different orientations of the proteins, the interface between the adsorbed protein layer and the solution is not located at a well defined distance from the substrate surface. Hence, there is no obvious location of the plane of the charge and the plane of the van der Waals force, which complicates the theoretical calculation of these forces. Due to these difficulties one can only draw firm conclusions based on large deviation of the measured forces from theoretical expectations.

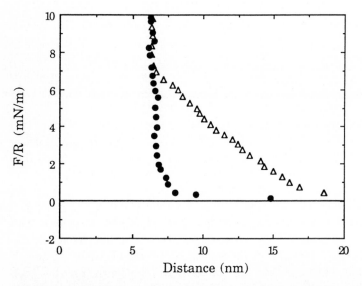

Fig. 8. Forces between mica surfaces across a 0.03 mg/ml lysozyme solution containing 1 mM NaCl, pH 5.6. The forces were measured before (●) and after (△) structural changes had been introduced in the layer by applying a large compressive force.

A second practical problem that has been observed when studying proteins on surfaces with the surface force technique is that irreversible changes in the adsorbed layer may take place when a strong compressive force is applied. This has to do with the fact that many proteins are neither very strongly or very weakly bound to the surface, causing them to be pushed out from between the surfaces only at so high forces that the surfaces have started to flatten (due to the deformation of the supporting glue). Under such conditions molecules at the edge of the flat region can leave the contact zone whereas those in the middle of the flat region will be trapped and pushed together (Blomberg et al. 1990). This phenomenon is related to elastohydrodynamic lubrication. When the surfaces are separated again, the "lump" of proteins that has formed in the middle of the contact region will remain at the contact position for a very long time, giving rise to long-range repulsive forces. A comparison between the forces observed before and after such pressure induced changes in the layer has occurred is seen in Fig. 8.

Hence, when studying proteins one should always determine how readily such changes in the adsorbed layer occur, and never apply such strong forces. As a comparison, weakly bound molecules will be removed from between the surfaces already when a weak force is applied (e.g. the outer protein layer of lysozyme in Fig. 7), whereas strongly bound molecules always will remain at the same position on the surface.

References

Åkesson T, Woodward C, Jönsson B (1989) Electrical double layer forces in the presence of polyelectrolytes. J Chem Phys 91:2461-2469

Berg JM, Claesson PM (1989) Forces between carboxylic acid layers in divalent salt solutions. Thin Solid Films 178:261-270

Berg JM, Claesson PM, Neuman R.D (in press) Interactions between mica surfaces in sodium polyacrylate solutions containing calcium ions. J Colloid Interface Sci.

Blomberg E, Claesson PM, Christenson HK (1990) Elastohydrodynamic effects with adsorbed layers in surface force measurements. J. Colloid Interface Sci 138:291-293

Blomberg E, Claesson PM, Gölander CG (1991) Adsorbed layers of human serum albumin investigated by the surface force technique. J. Dispersion Sci Techn 12:179-200

Chan D, Healy TW, White LR (1976) Electrical double layer interactions under regulation by surface ionization equilibrium - dissimilar amphoteric surfaces. J Chem Soc Faraday Trans I 72:2844-2865

Claesson PM, Berg JM (1989) The stability of carboxylic acid Langmuir-Blodgett films as studied by the surface force technique. Thin Solid Films 176:157-164

Claesson PM, Christenson HK (1988) Very long range attractive forces between uncharged hydrocarbon and fluorocarbon surfaces in water. J Phys Chem 92:1650-1655

Claesson PM, Blom CE, Herder PC, Ninham BW (1986a) Interactions between water-stable hydrophobic Langmuir-Blodgett monolayers on mica. J. Colloid Interface Sci 114:234-242

Claesson PM, Kjellander R, Stenius P, Christenson HK (1986b) Direct measurement of temperature-dependent interactions between non-ionic surfactant layers. J Chem Soc Faraday Trans I 82:2735-2746

Claesson PM, Herder PC, Blom CE, Ninham BW (1987) Interactions between a positively charged hydrophobic surface and a negatively charged bare mica surface. J Colloid Interface Sci 118:68-79

Claesson PM, Arnebrant T, Bergenståhl B, Nylander T (1989) Direct measurements of the interaction between layers of insulin adsorbed on hydrophobic surfaces. J Colloid Interface Sci, 130:457-466

Dahlgren MAG, Claesson PM, Audebert R (1993) Interaction and adsorption of polyelectrolytes on mica. Nordic Pulp and Paper Research Journal 8:62-67

Davies B, Ninham BW (1972) Van der Waals forces in electrolytes. J Chem Phys 56:5797-5801

Derjaguin BV (1934) Untersuchungen über die Reibung und Adhäsion IV. Kolloid Zeits 69:155-164

Herder CE (1991) Interaction between phosphine oxide surfactant layers adsorbed on hydrophobed mica. J Colloid Interface Sci 143:1-8

Israelachvili JN (1991) Intermolecular and surface forces - with applications to colloidal and biological systems, Academic Press, London, 2nd edition

Israelachvili JN, Adams GE (1978) Measurement of forces between two mica surfaces in aqueous electrolyte solutions in the range 0-100 nm. J Chem Soc Faraday Trans I 74:975-1001

Israelachvili JN, Pashley RM (1984) Measurement of the hydrophobic interaction between two hydrophobic surfaces in aqueous electrolyte solutions. J Colloid Interface Sci 98:500-514

Israelachvili JN, Wennerström H (1992) Entropic forces between amphiphilic surfaces in liquids. J Phys Chem 96:520-531

Malmsten M, Blomberg, E, Claesson PM, Carlstedt I, Ljusegren I (1992) Mucin layers on hydrophobic surfaces studied with ellipsometry and surface force measurements. J Colloid Interface Sci 151:579-590

Marra J, Israelachvili JN(1985) Direct measurements of forces between phosphatidylcholine and phosphatidylethanolamine bilayers in aqueous electrolyte solutions. Biochemistry 24:4608-4618

Parker JL., Attard P (1992) Deformation of surfaces due to surface forces. J Phys Chem 96:10398-10405

Parker JL., Claesson, PM (1994) Surface forces between hydrophobic silanated glass surfaces. Langmuir 10:635-639

Parker JL, Cho DL, Claesson PM (1989a) Plasma modification of mica: Forces between fluorocarbon surfaces in water and a nonpolar liquid. J Phys Chem 93:6121-6125

Parker JL, Christenson HK, Ninham BW(1989b) Device for measuring the force and separation between two surfaces down to molecular separations. Rev Sci Instrum 60:3135-3138

Parsegian VA., Gingell, D (1972) On the electrostatic interaction across a salt solution between 2 bodies bearing unequal charges. Biophysical J 12:1192-1204

Parsegian VA., Fuller N, Rand RP (1979) Measured work of deformation and repulsion of lecithin bilayers. Proc Natl Acad Sci USA 76:2750-2754

Pashley RM, McGuiggan PM, Ninham BW, Evans DF (1985) Attractive forces between uncharged hydrophobic surfaces: direct measurements in aqueous solutions. Science 229:1088-1089

Pezron I, Pezron E, Claesson PM, Bergenståhl B (1991) Monoglyceride surface films: Stability and interactions. J Colloid Interface Sci 144:449-457

Rabinovich YaI, Derjaguin, BV (1988) Interaction of hydrophobized filaments in aqueous electrolyte solutions. Colloids Surf 30:243-251

Tilton RD, Blomberg E, Claesson PM (1993) Effect of anionic surfactant on interactions between lysozyme layers adsorbed on mica. Langmuir 9:2102-2108

Tsao YH, Evans DF and Wennerström H (1993) Long-range attraction between a hydrophobic surface and a polar surface is stronger than that between two hydrophobic surfaces. Langmuir 9:779-785

Verwey EJN, Overbeek JThG (1948) Theory of the stability of lyophobic colloids, Elsevier, Amsterdam

3 Direct Measurements of Specific Ligand-Receptor Interactions Between Model Membrane Surfaces

J. Israelachvili, D. Leckband, F-J. Schmitt, J. Zasadzinski, S. Walker and S. Chiruvolu

3.1 Introduction

The Surface Forces Apparatus technique has made it possible to directly measure the long-range interaction forces and adhesion between two model membrane surfaces containing receptor and ligand molecules. Both long-range electrostatic and hydrophobic forces and short-range adhesion or specific binding forces can be directly measured at the ångstrom resolution level, and the rearrangements of the proteins and lipids during these interactions can also be studied. Results are presented of the measured forces between two surfaces, the one supporting a lipid-protein membrane exposing Streptavidin receptors, the other exposing Biotin ligands. At surface separations greater than 4Å three types of forces are operating : repulsive or attractive electrostatic forces and attractive van der Waals and hydrophobic forces. Closer in, a highly specific "lock and key" binding force suddenly switches on as soon as the surfaces approach within about 4Å of each other. The final binding is very strong, but the results show that the number of bonds formed, which determines the final adhesion strength, also depends on the fluidity of the supporting membranes and on the rates at which the ligands approach the surfaces. Our results also show which forces are responsible for different aspects of receptor-ligand interactions ; for example, while the longer-ranged electrostatic and hydrophobic forces are found to have little effect on the final adhesion energy, they do affect the rates of association and thereby play the dominant role in modulating the *on rates* of association in solution. Preliminary results are also presented on the experiments using biotin analogues having different binding constants to Streptavidin, and on parallel studies on specifically adhering vesicles in solution employing rapid freezing electron microscopy imaging techniques. These studies show that biospecific interactions such as are involved in immunological recognition and cell-cell contacts may be studied at the molecular level and in real time by the Surface Forces Apparatus (SFA) and Electron Cryo-Microscopy techniques.

3.2 Introduction

The physical forces between biological molecules and membranes are often highly specific, in contrast to non-specific interactions such as van der Waals and electrostatic (ionic) forces which are the major forces between colloidal particles and non-biological surfaces in solution, and which have been studied in great detail over the last 50 years (Israelachvili 1991 ; Chen et al. 1992). Here we describe a new method for the measurement of biospecific inter-membrane interactions where the full force-law (the force as a function of surface separation) can be directly determined and the molecular rearrangements during adhesion visualized.

3.3 Direct Force-Measuring Techniques

We used a Surface Forces Apparatus (SFA) to measure the forces and adhesion of various lipid-protein membranes supported on molecularly smooth solid surfaces (of mica). Details of the SFA technique have been fully described in the literature (Israelachvili 1991 ; Chen et al. 1992 ; Marra and Israelachvili 1985), but a brief account will be given here of how the SFA technique has been adapted for measuring biospecific interactions. In this technique, two thin sheets of atomically smooth mica are glued to the curved surfaces of two glass disks. The disks are then mounted into the apparatus facing each other. The separation D between the two surfaces is measured to within 1Å by viewing a coloured fringe pattern produced when white light is made to pass through them. These fringes also give the shapes of the two surfaces and any local deformations which can be continually monitored at the ångstrom level with a video camera during an interaction. The two surfaces can be made to approach or separate from each other with control to ±1Å using a variety of mechanical and piezoelectric distance controls, and forces are obtained from the deflection of a spring.

3.4 Preparation of Surfaces for Force Measurements

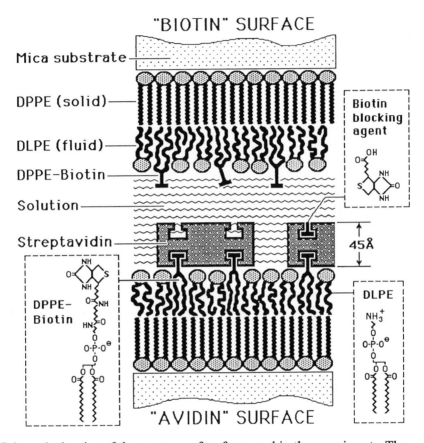

Fig. 1. Schematic drawing of the two types of surfaces used in the experiments. The Biotin surface exposes a fluid DLPE monolayer containing DPPE-Biotin ligand molecules at a surface coverage of 0.5 to 5%. The Avidin surface exposes the 45Å thick Streptavidin receptor proteins, each of which has four binding sites for biotin – two on either side of the molecule (Green 1975 ; Bayer and Wilchek 1990). **Materials:**   N-(6-((biotinyl)amino)hexanoyl)dipalmitoyl-L-α-phosphatidyl-ethanol amine (DPPE-Biotin), Dipalmitoylphosphatidyl-ethanolamine (DPPE), dilauryl-phosphatidyl-ethanolamine (DLPE). The Biotin Blocking Agent completely inactivates the site it binds to.

Two mica surfaces were coated with lipid bilayers whose outer fluid monolayers contained either the ligand L-α-dipalmitoyl-phosphatidyl ethanolamine-Biotin (DPPE-Biotin) or Streptavidin (the protein receptor) as shown in Fig. 1. For convenience and simplicity we shall call these surfaces the "Biotin" and "Avidin" surfaces. Streptavidin was obtained as a gift from Boehringer-Mannheim, lipids were from Sigma or Avanti Polar Lipids, and pure Biotin (blocking agent) was from Calbiochem. The Langmuir-Blodgett deposition technique (Marra and Israelachvili 1985) was used to prepare these surfaces. First, a primary close-packed monolayer of DPPE of area 0.42 nm^2 per molecule was deposited on a mica surface. This was followed by the deposition of a second fluid layer of L-α-dilaurylphosphatidyl-ethanolamine (DLPE) of area 0.55 nm^2 per molecule containing $0.5–5\%$ DPPE-Biotin (from Molecular Probes, Eugene, Oregon, USA), thereby exposing one Biotin ligand group per 1-10 nm^2. This deposition was done at $30°C$ from a 1 mM NaCl solution at a pressure of about 38 mN/m.

An "Avidin" surface was prepared from a "Biotin" surface by adsorbing soluble Streptavidin molecules (kindly provided by Boehringer-Mannheim GmbH, Werk Tutzing, FRG) from an aqueous solution, thereby yielding a surface with the same density of unsaturated Avidin (receptor) sites as Biotin (ligand) groups. Avidin surfaces could be prepared either before they were installed into the SFA apparatus or after installation inside the SFA chamber.

After the Biotin and/or Avidin surfaces were prepared, they were transferred (under water) into the SFA apparatus chamber previously filled with a 1 mM NaCl solution saturated with DLPE at the Critical Micelle Concentration. This was to ensure that no DLPE desorbed during the course of an experiment. The force-laws and adhesion energies were measured as previously described (Marra and Israelachvili 1985) over a range of temperatures from below to above the phase transition temperature of DLPE-bilayers ($30.5°C$). In addition, from the shapes of the optical interference (FECO) fringes used to "observe" the interacting surfaces one can directly measure not only the distance between two surfaces to ± 1Å, but also the diameter of the flattened contact area (to $\pm 1\mu$m) and other surface deformations associated with bilayer adhesion and fusion (Marra and Israelachvili 1985 ; Israelachvili 1991 ; Chen et al. 1992 ; Helm et al. 1992 ; Leckband et al. 1993)

3.5 Results

We found no unusual or specific interaction between two Avidin or two Biotin surfaces (Helm et al. 1991). In particular, the adhesion between these surfaces

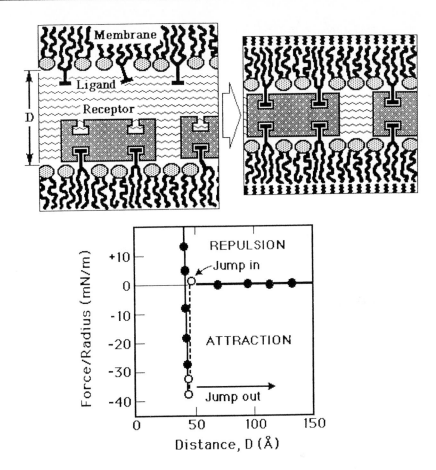

Fig. 2. Ligand-receptor system (*top*) as studied by the Surface Forces Apparatus technique and the resulting interaction potential (*bottom*) whose major feature is a very strong, short-range (<5Å) binding adhesion. Details of the full interaction potential are shown in Fig. 3.

was always weak and as expected for van der Waals forces. In contrast, an Avidin and a Biotin surface exhibited a very strong, short-range binding mechanism (Fig. 2) that is consistent with what is already known about the Avidin-Biotin system (Green 1975 ; Bayer and Wilchek 1990). A schematic of the full interaction potential for the Biotin-Avidin system is shown in Fig. 3, whose features will now be described.

3.5.1 Effects of pH and Salt Concentration

Changes in pH and ionic strength had no effect on the magnitude of the short-range adhesive force, suggestive of a non-electrostatic origin and consistent with a "lock and key" binding mechanism. However, they do affect the longer-ranged electrostatic "double-layer" force which determines the height of the "energy barrier" and thereby should affect the on-rates (kinetics of association) of ligands and receptors in solution.

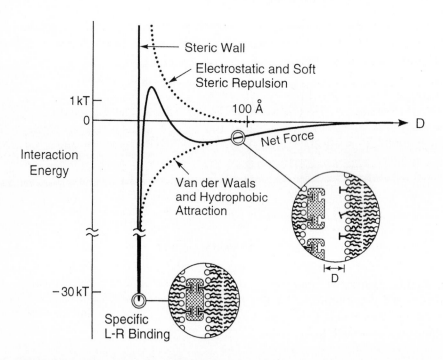

Fig. 3. Characteristic force-distance or energy-distance curve for ligand-receptor interactions. The various forces indicated can dominate at different distance regimes and under different solution conditions of electrolyte concentration, pH, temperature and approach rate, as explained in the text.

3.5.2 Effects of Temperature and Membrane Fluidity

The effect of receptor-ligand mobility on the membrane-membrane adhesion was studied by measuring how the short-range adhesion force changed when the temperature was raised from below to above the chain melting temperature (T_c~30°C) of the lipids in the supporting bilayers. Increasing the temperature from 25°C to 33°C was found to increase the adhesion by seven-fold relative to frozen bilayers at $T<T_c$. This was attributed to the enhanced diffusion and rates of molecular rearrangement in the more fluid bilayers which resulted in greater and more rapid intermembrane bond formation. These results indicate that receptor lateral mobility in membranes is a critical determinant of the strength of specific intermembrane adhesion.

3.5.3 Steric and Time-Dependent Interactions

The interaction potential shown in Fig. 3 depended on the rate at which the two surfaces were brought towards each other. If the surfaces of two *fluid* membranes were brought together quickly, an enhanced short-range repulsion (higher energy barrier) was measured which, however, disappeared with time or if the surfaces were made to approach more slowly. This short-range repulsion is attributed to the steric effects produced by the protruding, polymer-like Biotin groups as they bump into the opposite surface before they find the receptor pocket and lock into it. This type of thermal or soft steric repulsion is in contrast to the stronger and sharper steric repulsion (steric wall at 45Å in Figs. 2 and 3) that occurs once the molecules of the two surfaces are fully in contact with each other. A further indication that the soft steric repulsion is due to the non-equilibrium interactions of protruding surface groups was the observation that with *frozen* membranes (at $T<T_c$) this repulsion did not go away with time, exactly as expected if the membrane components were no longer free to freely diffuse in the plane of the membranes (Leckband et al. 1992).

3.5.4 Effect of Ligand and Receptor Surface Concentration

The intermembrane binding strength was also measured as a function of the biotin surface concentration, with both fluid and frozen bilayers. Binding strengths were measured at two biotin-lipid concentrations of 5% and 0.5% which differ by an order of magnitude. At 25°C, the tenfold reduction in the biotin coverage resulted in an 82% reduction in the membrane adhesion compared to the theoretically expected decrease of 80% assuming a linear relation between binding and surface concentration. In contrast, at 33°C, the adhesion decreased by 60%. Overall, it appears that intersurface adhesion varies almost linearly with the ligand surface density.

3.5.5 Effect of Receptor Blocking Agents

Experiments were conducted with soluble Biotin used as a blocking agent (Fig. 1) which bound tightly to the receptor sites and inactivated the Streptavidin molecules. As expected, the forces between Biotin surfaces and inactivated Avidin surfaces no longer exhibited the strong, short-range binding attraction ; but the longer-ranged forces were also found to be affected. The results indicate that an additional medium-to-long range attractive force operates between Biotin ligands and active Streptavidin receptors. In view of recent measurements of similar forces between stressed lipid bilayer systems (Helm et al. 1992), it is tentatively concluded that this additional force is due to a hydrophobic interaction between the Biotin group and the receptor site (when active), which is switched off once the site is occupied or blocked.

3.6 Parallel Studies of Interacting Vesicles in Solution

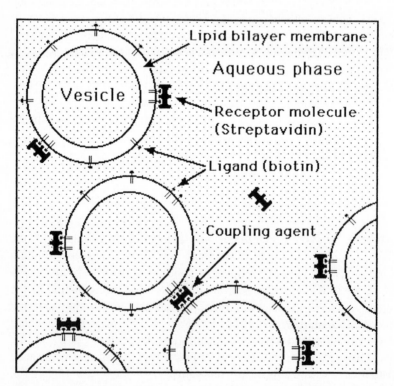

Fig. 4. Biotinylated vesicles in solution being bound together by Streptavidin.

Experiments have also been carried out on interacting vesicles in solution. One of the aims was to study the difference between non-specific and site specific adhesion. Single bilayer vesicles of lecithin were prepared as previously described (Bailey et al. 1990), but with the outer surfaces containing biotinylated lipids that exposed biotin ligands to the aqueous phase (Fig. 4).

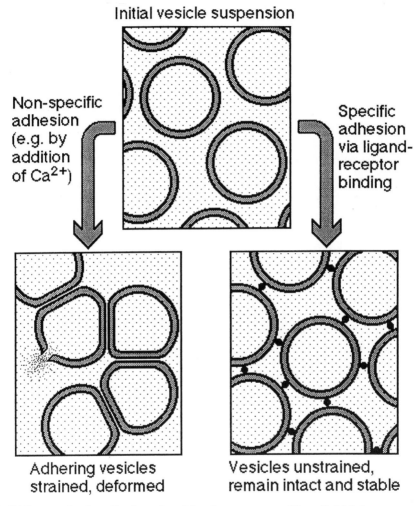

Fig. 5. Difference in the adhesion of vesicles due to nonspecific colloidal forces (*left*) and site-specific binding (*right*).

Addition of soluble Streptavidin receptors to the solution caused the vesicles to aggregate without deforming them. This was brought about by the simultaneous binding of one Streptavidin molecule to ligands on two different vesicles surfaces which crosslinked or tethered them together.

This mode of binding contrasts sharply with that produced by the non-specific aggregation of vesicles (Fig. 5) brought about by changing the conventional colloidal forces between them, for example, via addition of Ca^{2+} to a medium containing negatively charged lipid vesicles (ion binding aggregation), addition of PEG to a suspension of uncharged vesicles (depletion aggregation), laterally stressing or osmotically swelling vesicles (osmotic stress or hydrophobic aggregation). In each of these aggregation mechanisms, the adhesion energy of the *whole* vesicle surface is increased, so that the resulting interaction between vesicles causes them to deform by flattening because the surfaces want to increase their contact area as much as possible (Fig. 5 left). These deformations are known to cause large elastic and osmotic stresses on the membranes which can result in increased fragility, leakage, rupture and/or fusion. The stronger the adhesion the larger are these effects so that it is generally not possible to increase inter-vesicle or inter-particle adhesion without at the same time weakening the strength (decreasing the stability) of the aggregate. In contrast, no such deformations occur when vesicles are brought together via a site-binding interaction because the whole surface is not involved – only the highly localized ligand groups on the surfaces are involved. The result is that strong inter-vesicle or intermembrane binding is achieved without at the same time stressing or perturbing the membranes (Fig. 5 right).

3.7 Discussion and Conclusions

These direct force measurements demonstrate that it is possible to directly study highly specific biomolecular interactions at the molecular level and in "real time".

The results on the Streptavidin-Biotin system show that in addition to the expected attractive van der Waals and repulsive electrostatic forces, there is also a long-range attractive, electrostatic or hydrophobic, force between the active receptor and ligand moieties that most likely play a significant role in steering bimolecular trajectories. This attractive force is opposed by short-range electrostatic and steric forces, and together they give rise to an energy barrier at some finite separation that does not affect the final strong adhesion (binding) energy but would have a marked effect on the association kinetics or on-rates at which molecules bind (Leckband et al. 1992, 1994).

Intermembrane interactions were also shown to be modulated by the fluidity of the supporting membranes and by the receptor and ligand densities. The increased membrane fluidity correlated directly with a stronger intermembrane adhesion, presumably because the increased lateral mobility and greater ease of molecular reorientations result in more bonds being formed per unit time.

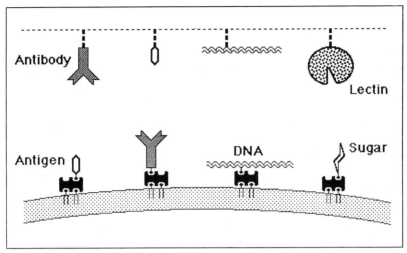

Fig. 6. Model-membrane surfaces that can now be prepared for detailed study with the Surface Forces Apparatus technique as well as with labelled vesicles in solution which together provide excellent opportunities for studying the interactions of biological macromolecules, biomembranes and model cell surfaces.

It is clear that nature has here developed a very efficient mechanism by which noncovalent adhesive junctions having the strength of covalent bonds can be switched on, or mechanically "locked", quickly and with minimal expenditure of energy. The overall mechanism is, however, complex and multi-facetted, with different parts of the interaction potential being affected or controlled by different conditions or stimuli which, in turn, regulate different stages of the binding process. As previously discussed in the literature (Evans 1985a, b), discrete site-binding processes can be very different and much more complex than smeared out (continuum type) adhesive junctions, and it is highly unlikely that a single binding constant or association constant could ever describe all the subtle static and dynamic features of these interactions.

3.8 Future Studies

Our results have important consequences for understanding the role of cell surface dynamics in cell attachment processes and the resulting adhesion strength. The successful application of the SFA technique to the study of the Avidin-Biotin system opens the way for studying other types of similar interactions, some of which are illustrated in Fig. 6. Experiments are also progress (Leckband et al. in preparation) to study how the interactions of Biotin analogues, having different binding affinities for Avidin, contrast with the overall interaction of the system described here which, with a binding constant of 10^{-15} M, is one of the highest known.

Acknowledgements
This work was supported by a grant from the National Institutes of Health (PHS GM47334) and made use of MRL Central Facilities supported by the National Science Foundation (DMR-9123048).

References

Bailey SM, Chiruvolu S, Israelachvili JN, Zasadzinski JAN (1990) Measurements of forces involved in vesicle adhesion using freeze fracture electron microscopy. Langmuir 6:1326-1329

Bayer E, Wilchek M (1990) Biotin-binding proteins : overview and prospects. Methods in Enzymology 184:49-51

Chen YL, Kuhl T, Israelachvili JN (1992) Mechanism of cavitation damage in thin liquid films : collapse damage vs. inception damage. Wear 153:31-51

Evans EA (1985a) Detailed mechanics of membrane-membrane adhesion and separation. I. Continuum of molecular cross-bridges. Biophys J 48:175-183

Evans EA (1985b) Detailed mechanics of membrane-membrane adhesion and separation. II. Discrete kinetically trapped molecular cross-bridges. Biophys J 48:185-192

Green M (1975) Avidin. Adv Protein Chem 29:85-133

Helm C, Knoll W, Israelachvili JN (1991) Measurement of ligand-receptor interactions. Proc Natl Acad Sci USA 88:8169-8173

Helm CA, Israelachvili, JN, McGuiggan PM (1992) Role of hydrophobic forces in bilayer adhesion and fusion. Biochemistry 31:1794-1805

Israelachvili J (1991) Intermolecular and surface forces. Academic Press, New York.

Leckband DE, Schmitt F J (submitted).

Leckband DE, Israelachvili JN, Schmitt FJ, Knoll W (1992) Long-range attraction and molecular rearrangements in receptor-ligand interactions. Science 255:1419-1421

Leckband DE, Helm CA, Israelachvili JN (1993) Role of calcium in the adhesion and fusion of bilayers. Biochemistry 32:1127-1140

Leckband DE, Schmitt FJ, Israelachvili JN, Knoll W (1994) Direct force measurements of specific and nonspecific protein interactions. Biochemistry 33:4611-4624

Marra J, Israelachvili J. (1985) Direct measurement of forces between phosphatidylcholine and phosphatidylethanolamine bilayers in aqueous electrolyte solutions. Biochemistry 24:4608-4618

4 The Study of Biomolecules on Surfaces Using the Scanning Force Microscope

R. Erlandsson and L. Olsson

4.1 Introduction

The invention of the Scanning Tunnelling Microscope in 1981 (Binnig et al. 1982) started the development of a whole class of new techniques, all based on probing surface properties by laterally scanning a sharp tip in the proximity of a surface. One of the most promising of these methods for biological applications is Scanning Force Microscopy, SFM (also known as Atomic Force Microscopy, AFM ; Binnig et al. 1986). This technique, which has been commercially available for several years[1] has some specific features that make it attractive for biological measurements :

- It has potentially very high resolution (sub nm range)
- It can image non-conducting species and does not need conducting substrates
- It can operate in liquids like physiological buffer

For inorganic systems, atomic resolution has been obtained under favourable conditions but on soft biological materials, the lateral resolution is at present typically in the 1-40 nm range and depends strongly on the sample. While the SFM technique is primarily an imaging method, giving the three-dimensional topography of the sample, there are also examples of other ways of using it. It has been demonstrated how molecules and membranes can be manipulated by the probe tip (Henderson 1992 ; Hoh et al. 1992) and the technique has also been used to locally measure properties like surface forces (Ducker et al. 1991 ; Hoh et al. 1992).

---

[1]Digital Instruments Inc., 520 E. Montecito St., Santa Barbara, CA 93103, USA ; Park Scientific Instruments, 1171 Borregas Ave, CA 94089, USA ; TopoMetrix, 1505 Wyatt Drive, Santa Clara CA 95054, USA.

4.2 The SFM Technique

Figure 1 shows a schematic view of a scanning force microscope using a fiber interferometer to detect the lever motion. The sample is raster scanned in the x-y plane with respect to the probe tip using a piezoelectric device while measuring the force interaction between the tip and the sample. The force signal is used as input for a feedback circuit that controls the z-position of the sample in order to keep the force constant. Plotting z=f(x,y) will give a topographic representation of the sample surface. In order to measure the force, the tip is mounted at the end of a small flexible element, the lever. In the basic mode of operation, the deflection of the lever directly gives the force on the tip. To measure the

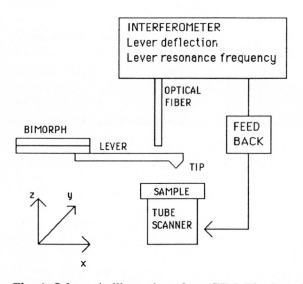

Fig. 1. Schematic illustration of our SFM. The laser interferometer is used to measure either lever deflection or the shift in resonance frequency. See text for further details.

lever deflection, tunnelling was originally used, but modern instruments use optical techniques like beam deflection (Meyer and Amer 1988) or interferometry (Rugar et al. 1988). The tip and the lever are normally built as a unit, microfabricated out of silicon or silicon nitride. It is sometimes preferable to use metallic tips which normally are made by electrochemically etching and bending a tungsten wire. The instrument can be operated in several different modes that will be discussed below.

4.2.1 Forces Acting on the Probe Tip

When the probe tip approaches a surface in vacuum, the attractive van der Waals force will cause the tip to jump into repulsive contact at a critical separation where the spatial derivative of the attractive force equals the lever spring constant. If the measurement is taking place in air, there will be an additional attractive interaction as the tip is trapped in a thin liquid layer on the surface that will give rise to a capillary force (Grigg et al. 1992a). This liquid contact typically occurs at a separation of 5-10 nm, but it can vary significantly with sample and environment. If the lever has a low spring constant, the capillary force will pull the tip into repulsive contact with the surface, while a higher spring constant can lead to a situation with a liquid bridge in the tip-surface gap. When the separation has reached this 'instability point' where the attractive forces have pulled the tip in hard contact with the surface, the force on the tip is a complex combination of a repulsive component over a small area of the tip, an attractive van der Waals force acting over a larger area, an attractive capillary force and the lever spring force.

4.2.2 Contact Mode SFM

This is the basic measurement mode of the force microscope that has been used in most of the measurements presented in the literature. The tip is in contact with the sample and subject to repulsive as well as attractive forces from the surface as described above. The feedback keeps the lever deflection at a pre-set value during the measurement and the contact force can be adjusted by varying this value. If the measurement is performed in air, the dominating attractive force is from the capillary interaction which will set a practical limit on how low the contact force can be adjusted. A typical value in air is 50 nN.

By enclosing the sample and lever in a sealed cell, the contact mode technique is easily adapted for measurement in liquids. Liquid phase measurement makes it possible to study biological systems in their native environment, and has the additional advantage that it eliminates the capillary force, making it possible to track the surface using much lower contact force than in air. The contact force in the liquid phase can be well below 1 nN, in some cases. Values down to 0.01 nN have been reported.

The resolution in contact mode operation is defined by the repulsive contact area and can be very high (0.1 nm laterally, 0.01 nm in height) for some flat and hard inorganic systems. For soft samples, like biomolecules, the highest lateral resolution is on the order of 1 nm, but varies within a wide range depending on the sample.

4.2.3 Non-Contact Mode SFM

In this mode of operation the tip is not allowed to come into repulsive contact with the sample, and the separation is regulated by measuring the attractive force. Because of the weak interaction in this regime, it is advantageous to use a resonance method (Marti et al. 1987) to detect the proximity of the surface. In this case, the lever is oscillated close to its resonance frequency with an amplitude of typically 1 nm using a piezoelectric element (the bimorph in Fig. 1). The resonance frequency is then modified by the derivative of the force between the surface and the tip, F', according to:

$$\frac{d\omega}{\omega} = \frac{F'}{2k} \tag{1}$$

where ω is the undisturbed resonance frequency of the lever, k is the lever spring constant and $d\omega$ is the induced frequency shift. The frequency shift can be measured by detecting how the oscillation amplitude varies provided that the resonance curve is known and no damping occurs. If the separation is reached where a liquid contact takes place, there will be a discontinuous drop in the amplitude. Figure 2 shows how the lever deflection (upper curve) and resonance amplitude (lower curve) varies as a function of sample position for a tungsten tip as it first approaches and then is retracted from a SiO_2 surface. Point A in the figure indicates the formation of a liquid bridge, point B indicates hard contact and point C indicates where the liquid contact is broken as the tip is retracted. It is seen from the figure that there is a continuous variation of the resonance amplitude both before liquid bridging (directly related to the force derivative) and after the liquid bridge formation (corresponding to a separation dependent damping). As the resonance amplitude is used as the input for the feedback circuit, it is consequently possible to operate the instrument both with and without a liquid bridge by using the appropriate regulation point. The non-contact mode will not give the same ultimate resolution as contact mode measurements have demonstrated on hard samples, but has the advantage that artefacts due to tip forces are minimized. Non-contact mode force microscopy has mainly been used in studies of electrostatic and magnetic effects but has also been successfully applied to investigations of biological macromolecules, as demonstrated by measurements on fibrinogen adsorption on SiO_2 done at our laboratory (Wigren et al. 1991) showing a lateral resolution of better than 5 nm.

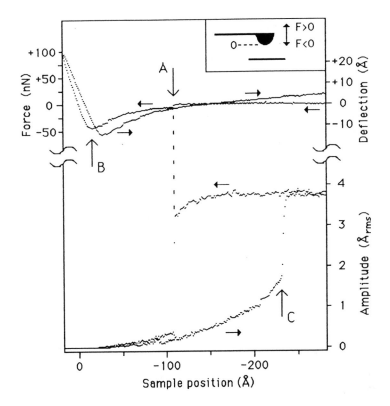

Fig. 2. See text for further details.

4.2.4 Tapping Mode SFM

An additional mode of operation was recently introduced by Digital Instruments   called the "Tapping mode" SFM. As in the non-contact mode, the lever is oscillated close to the resonance frequency, but with a much higher amplitude (up to 100 nm) so that the tip strikes the surface and is intermittently trapped in the liquid layer for each oscillation cycle. As the damping of the oscillation is strongly dependent on the separation, it can be used as the input signal for the feedback system. Even though there is a repulsive contact during part of each oscillation cycle, it is claimed that the force per strike is 0.1-1 nN, which is less than for contact mode measurements in air.

4.3 Imaging Biomolecules

It was early demonstrated that force microscopy could be used to image certain organic systems with molecular resolution such as crystals of amino acids (Drake et al. 1989). High resolution data showing the molecular organization of Langmuir-Blodgett (LB) films, that can serve as models for biological membranes, were also reported (Meyer et al. 1991). The high (sub nm) resolution obtained in these cases is possible as the structures are ordered, making them more resistant to damage due to the tip force. For biological systems, the highest resolution has also been obtained for ordered structures such as supported planar membranes with bound proteins (Weisenhorn et al. 1990a) and certain native membranes [purple membranes (Butt et al. 1990), gap junctions (Hoh et al. 1991)]. Also individual proteins have been imaged, like immunoglobulins, phosphorylase, fibrinogen and fibronectin (Wigren et al. 1991 ; Drake et al. 1989 ; Edstrom et al. 1990 ; Emch et al. 1992), but with lower resolution (typically 5 nm laterally). Imaging DNA has attracted a lot of interest (Henderson 1992 ; Vesenka et al. 1992 ; Hansma et al. 1992 ; Thundat et al. 1992), and it is now possible to routinely obtain stable images of plasmids. An interesting aspect of the SFM technique is the possibility to follow processes in real time, as was demonstrated in reference by Drake et al. (1989), where the polymerization of fibrin was recorded after thrombin was added to a fibrinogen solution.

4.3.1 Image Artefacts

Obtaining reproducible and reliable images of biomolecules is often a challenge as they are fragile and often adhere poorly to the substrate. The two main problems that affect most SFM measurements of these materials are:

- Distortion and damage due to tip forces
- Artefacts due to the finite tip geometry

The most commonly observed effect due to tip loading forces is displacement of molecules, giving poor reproducibility between scans. It has also been observed how the tip acts as a "molecular broom" that orients protein molecules in strands across the image (Lea et al. 1992). These kinds of effects are often apparent if the scan area is increased, making the modifications in the previously scanned area visible. If a molecule sticks well enough to the surface not to be displaced, an increase of the loading force will normally decrease the apparent height and broaden the structure. This effect was shown by Bustamante et al. (1992) for a plasmid DNA molecule that was imaged with loading forces in the range of 20-170 nN in air. At 110 nN the molecule began to appear distorted, but the effect was reversible, while a force of 170 nN created irreversible damage.

For measurements in air, it is well established that image artefacts can be directly related to high humidity (Thundat et al. 1992), which, at least partly, is a consequence of increased tip load due to stronger capillary force. As the loading force increases, the lateral friction force on the tip will also increase which can lead to a twisting of the lever that will be interpreted as a topographic change by the control system. This has been observed in measurements of DNA in air (Thundat et al. 1992), where the contrast was inverted for structures imaged at high humidity. As the humidity increased, the capillary force increased preferentially on the DNA molecules, leading to stronger lateral force in these areas.

To minimize the problems described above, the instrument should be operated with lowest possible contact force. For contact mode measurements, the best way to decrease the necessary tip load is to operate under liquid, eliminating the problems due to capillary forces. Significant forces can, however, also occur during measurements in liquids due to surface charge and hydrophobic interaction, but these effects can be minimized by the right combination of liquid and substrate (Ohinishi et al. 1992). For measurements in air it can be an advantage to use tips with a small cone angle as the area on the tip contributing to attractive interactions decreases relative to the repulsive contact zone active in imaging. Working in the non-contact mode radically decreases the contact force, but as this operation mode has not been available on commercial instrumentation until recently it has not been much used.

Artefacts due to the finite tip size are an intrinsic problem for the technique and can cause several types of distortions in the images. Typical signs to look out for are broadenings, frequently reoccurring shapes that vary with scan rotation and the existence of shadows and double structures. The standard micro-fabricated tips are pyramidal and have a tip radius of 10-40 nm, but today there are also improved versions available with higher aspect ratios. One way of obtaining tips with extremely low cone angles is to deposit a needle like carbon "microtip" on top of a standard pyramidal tip, which can be done using an electron microscope (Keller and Chih-Chung 1992). The recorded data will always be a convolution of the tip shape and the sample shape, but the degree of the distortion varies with the relative size between the tip apex and the imaged object. The ultra high (atomic) resolution that can be obtained on atomically flat inorganic materials is a consequence of an atomically sharp micro contact between an asperity on the tip ("microtip") and the surface. For the measurements discussed here, the situation is radically different, as the size of the structures is often comparable with the tip radius. For structures taller than the tip radius, the cone angle of the tip will be the main factor determining the observed broadening, while for smaller structures the tip radius is the relevant parameter. From a simple geometric argument it can be shown that for ideal shapes, the measured width of a cylindrical molecule with a radius much smaller than the tip radius will be (Vesenka et al. 1992) :

$$W = 4\sqrt{R_t R_m} \tag{2}$$

where R_t is the tip radius, R_m the molecule radius and W the measured width. In reality, the tip surface is rough, which means that the contact point can change as the tip passes over the molecule, causing distortion. If more than one contact point is active at the same time, characteristic double structures or shadow effects will occur. By obtaining images from a known structure like a sharp step, it is possible to estimate the tip shape, which can be used to obtain the true width of the observed structures (Grigg et al. 1992b).

4.3.2 Sample Preparation

One of the most attractive features of the SFM technique is the ability to work in liquid and image biomolecules in their native state, without need for coating, freeze drying etc. that are needed to preserve the samples using electron beam based techniques. Even though air drying can be a very damaging process, there are also many examples of biological systems, ranging from individual protein molecules to whole cells, that have been studied with SFM at ambient pressure. The most important requirements on the substrate materials, whether the measurement takes place in air or liquid, are that they should anchor the observed species firmly enough to avoid the force induced artefacts discussed above, and that the surfaces should be flat on the scale of the observed structures.

There exists no simple recipe that is generally applicable to achieve sufficient adhesion, but a multitude of different approaches that often are specific to the chemistry of the studied system. Some general aspects are given below : the most used substrate materials are mica and glass, but also HOPG graphite, Au and SiO_2 are sometimes used. The mica surface is flat on the molecular scale and has normally a negative surface charge that can give a sufficient electrostatic attraction to attach positively charged species. If a positively charged surface is needed, the mica can be modified by exposure to Mg^{2+} before adsorbing the sample. To prepare samples for air measurements, the substrate is normally incubated in a buffer with the sample molecules, rinsed in distilled water and blown dry in nitrogen. In some cases, precipitation of salts on the substrate can be a problem, especially if rinsing in water is not desirable. Using a volatile buffer like ammonium acetate can be a solution in these cases (Emch et al. 1992).

More elaborate methods are also used to obtain the desired adhesion properties like chemically altering the substrate surfaces, providing an intermediate layer with properties that can be tailored to a specific molecule. Silanization of glass and SiO_2 can be used to change surface properties like surface energy but can also be used to covalently link specific samples to the surface (Ohnesorge et al. 1992). Chemical modification involving covalent attachment of chemically polarizable groups by reactions of thiols and disulfides with gold surfaces has been used to immobilize biomolecules (Bottomley et al. 1992). Polymerized Langmuir-Blodgett films have also been used to covalently bind DNA molecules (Weisenhorn et al. 1990b). When imaging molecules under liquids that are electrostatically attached to the substrate, it has been observed that non-polar liquids like propanol with low dielectric constant give a better adherence (Hansma et al. 1992) than water or physiological buffers, as they do not screen the electrostatic forces holding down the molecules.

4.4 Non-Contact Imaging of Fibrinogen

As an example of non-contact imaging of biological molecules, we present here a measurement of fibrinogen adsorption on SiO_2 surfaces with varying hydrophobicity. As substrates we used pieces of polished silicon wafers with native oxide which first had been cleaned in a mixture of hot (80°C) $H_2O/H_2O_2/NH_4$ (5:1:1) rinsed in deionized water, then cleaned in hot (80°C) $H_2O/H_2O_2/HCl$ (6:1:1) and followed by extensive rinsing in deionized water. This treatment gives an oxidised and hydrophilic silicon surface. The Si-pieces were made hydrophobic by immersing them into 10% dichloro-dimethylsilane in trichlorethylene for 5 min followed by rinse in trichloroethylene and ethanol. Fibrinogen molecules were adsorbed on the treated Si surfaces by incubation at different concentrations at room temperature with slow stirring. The surfaces were rinsed in distilled water

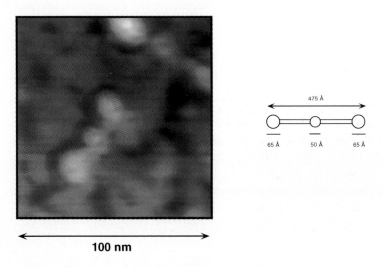

Fig. 3. Fibrinogen on hydrophilic silicon (10 µg/ml, 10 min).

and blown dry in clean nitrogen before they were mounted in the microscope which was operated in ambient atmosphere.

The instrument was operated in the non-contact mode described above, using a tungsten tip/lever unit that had been electrochemically etched from a tungsten wire. Figure 3 shows a top-view representation where two individual fibrinogen molecules can be seen, both showing the characteristic tri-nodular shape that has also been observed by TEM (Nygren and Stenberg 1988). By changing the feedback parameters, this molecule could be resolved both with and without a liquid contact between tip and sample. The resolution is remarkably high for non-contact imaging and the molecules could be scanned for long periods without observing any change, which demonstrates that the tip-sample force is to small to disturb the molecules. For low concentrations, the characteristic tri-nodular shaped molecules are observed both on the hydrophobic (Wigren et al. 1987) and hydrophilic surfaces. For higher exposures, however, there is a significant difference in molecule distribution. On the hydrophilic surface the density of adsorbed molecules increases with exposure without any specific rearrangement. On the hydrophobic surface, on the other hand, we observe pronounced clustering of the molecules and eventually formation of a dense network indicating a stronger inter-molecular interaction in this case, see Fig. 4. A similar trend was observed by Emch et al. (1992) for

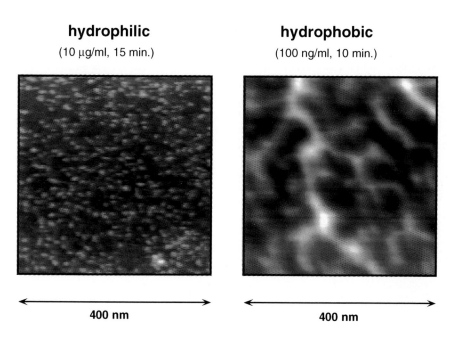

hydrophilic
(10 µg/ml, 15 min.)

hydrophobic
(100 ng/ml, 10 min.)

400 nm

400 nm

fibronectin adsorbed on mica (hydrophilic) and PMMA (hydrophobic), where network formation only occurred on the PMMA surface.

References

Binnig G, Rohrer H, Gerber C, Weibel E (1982) Surface studies by scanning tunneling microscopy. Phys Rev Lett 49:57-61

Binnig G, Quate C, Gerber C (1986) Atomic force microscope. Phys Rev Lett 56:930-933

Bottomley LA, Haseltine JN, Allison DP, Warmack RJ, Thundat T, Sachleben RA, Brown GM, Woychik RP, Jacobson KB, Ferrell Tl, Lucovsky G (1992) Scanning Tunneling microscopy of DNA : the chemical modification of gold surfaces for immobilization of DNA. J Vac Sci Technol A 10(4):591-595

Bustamante C, Vesenka J, Tang CL, Rees W, Guthold M, Keller R (1992) Circular DNA molecules imaged in air by scanning force microscopy. Biochemistry 31:22-26

Butt HJ, Dowing KH, Hansma PK (1990) Imaging the membrane protein bacteriorhodopsin with the atomic force microscope. Biophys J 58:1473-1480

Drake B, Prater CB, Weisenhorn AL, Gould SAC, Albrecht TR, Quate CF, Cannell DS, Hansma HG, Hansma PK (1989) Imaging crystals, polymers, and processes in water with the atomic force microscope. Science 243:1586-1589

Ducker WA, Senden TJ, Pashley RM (1991) Direct measurement of colloidal forces using an atomic force microscope.Nature 353:239-241

Edstrom RD, Meinke MH, Yang X, Yang R, Elings V, Evans F (1990) Direct visualization of phosphorylase-phosphorylase kinase complexes by scanning tunneling and atomic force microscopy. Biophys J 58:1437-1448

Emch R, Zenhausern F, Jobin M, Taborelli M, Descouts P (1992) Morphological difference between fibronectin sprayed on mica and on PMMA. Ultramicroscopy 42-44:1155-1160

Grigg DA, Russell PE, Griffith JE, Lucovsky G (1992a) Tip-sample forces in scanning probe microscopy in air and vacuum. J Vac Sci Technol A10(4):680-683

Grigg DA, Russell PE, Griffith JE, Vasile MJ, Fitzgerald EA (1992b) Probe characterization for scanning probe metrology. Ultramicroscopy 42-44:1616-1620

Hansma HG, Vesenka J, Siegerist C, Kelderman G, Morett H, Sinsheimer RL, Elings V, Bustamante C, Hansma PK (1992) Reproducible imaging and dissection of plasmid DNA under liquid with the atomic force microscope. Science 256:1180-1184

Henderson EH (1992) Imaging and nanodissection of individual supercoiled plasmids by atomic force microscopy. Nucleic Acid Res 20:445-447

Hoh JH, Cleveland JP, Prater CB, Revel JP, Hansma PK (1992) Quantized adhesion detected with the atomic force microscope. J Am Chem Soc 114:4917-4918

Hoh JH, Lal R, John SA, Revel JP, Arnsdorf MF (1991) Atomic force microscopy and dissection of gap junctions. Science 253:1405-1408

Hoh JH, Revel JP, Hansma PK (1992) Membrane-membrane and membrane-substrate adhesion during dissection of gap junctions with the atomic force microscope. Proc Soc Photo-Opt Instrument Eng 1639:212-215

Keller D, Chih-Chung C (1992) Imaging steep, high structures by scanning force microscopy with electron beam deposited tips. Surf Sci 268-333

Lea AS, Pungor A, Hlady V, Andrade JD, Herron JN, Woss EW Jr. (1992) Manipulation of proteins on mica by atomic force microscopy. Langmuir 8:68-73

Marti Y, Williams CC, Wickramasinghe HK (1987) Atomic force microscope-force mapping and profiling on a sub 100-Å scale. J Appl Phys 61:4723-4729

Meyer E, Howard L, Overney RM, Heinzelmann H, Frommer J, Günterherodt HJ, Wagner T, Schier H, Roth S (1991) Molecular-resolution images of Langmuir-Blodgett films using atomic force microscopy. Nature 349:398-400

Meyer G, Amer NM (1988)   Novel optical approach to atomic force microscopy. Appl Phys Lett 53:1045-1047

Nygren H, Stenberg J (1988)   Molecular and supramolecular structure of adsorbed fibrinogen and adsorption isotherms of fibrinogen at quartz surfaces. Mater Res 22 :1-11

Ohnesorge F et al. (1992) Scanning force microscopy studies of the S-layers from Bacillus coagulans E38-66, Bacillus sphaericus CCM2177 and of an anbitody binding process. Ultramicroscopy 42-44:1236-1242

Ohinishi S, Hara M, Furuno T, Sasabe H (1992) Imaging the ordered arrays of water-soluble protein ferritin with the atomic force microscope. Biophys J 63:1425-1431

Rugar D, Mamin HJ, Erlandsson R, Stern JE, Terris BD (1988) Force microscope using a fiber-optic displacement sensor. Rev Sci Instr 59:2337-2340

Thundat T, Allison DP, Warmack RJ, Ferrell TL (1992) Imaging isolated strands of DNA molecules by atomic force microscopy. Ultramicroscopy 42-44:1101-1106.

Vesenka J, Guthold M, Tang CL, Keller D, Delaine E, Bustamante C (1992) Substrate preparation for reliable imaging of DNA molecules with the scanning force microscope. Ultramicroscopy 42-44:1243-1249

Weisenhorn A, Drake B, Prater CB, Gould SAC, Hansma PK, Ohnesorge F, Egger M, Heyn SP, Gaub HE (1990a) Immobilized proteins in buffer imaged at molecular resolution by atomic force microscopy. Biophys J 58:1251-1258

Weisenhorn A, Gaub HE, Hansma HG, Sinsheimer RL, Kelderman GL, Hansma PK (1990b) Imaging single-stranded DNA, antibody-antigen reaction and polymerized Langmuir-Blodgett films with an Atomic Force Microscope. Scanning Microscopy 4:511-516

Wigren R, Elwing H, Erlandsson R, Welin S, Lundström I (1991) Structure of adsorbed fibrinogen obtained by scanning force microscopy. FEBS Lett. 280:225-228

5 Surface Force Measurement Techniques and Their Application to the Interaction Between Surfactant Coated Surfaces

J. L. Parker

5.1 Introduction

When any two interfaces are brought together they will exert a force upon one another before they contact. There is a diverse number of different types of interactions which make up the total force between two solid surfaces and may include contributions from electrostatic, van der Waals, solvation, core exclusion, steric, "hydrophobic", and correlation interactions as well as many others (Israelachvili 1991). The range at which the force is felt depends on the type of interaction and this is dependent on both the surfaces and the medium. Electrostatic forces can be felt micrometers from the surface whereas solvation forces are only encountered in the last few molecular layers between two surfaces.

A surface force apparatus is a device which can be used to directly measure interactions between solid surfaces. The data produced by such devices are the force as a function of the separation between two surfaces. These data can be compared with theory and the nature of the interactions between the surfaces deduced. For instance, for two equally charged surfaces the interaction at large separations is usually well described by DLVO theory, and from fits of this theory to the data basic information such as the charging properties of the surfaces can be obtained.

Even with accurate information on both the structure of the interacting surfaces and the physical properties of the medium in which they interact, it is often impossible to predict a priori the force law between the surfaces. For instance, on the basis of theory, one would expect that the interaction between two macroscopic hydrophobic surfaces would be attractive with a power law dependence typical of a van der Waals force: $F/R = -A/6D^2$. A reasonable value for the Hamaker constant (A) would be 1×10^{-20} J . This is not what is observed, instead an attractive interaction much stronger than the van der Waals interaction is measurable to much larger separations. Furthermore, the force law has an exponential form rather than a power law (Claesson al. 1986 ; Christenson and Claesson 1988). Many other examples exist where unexpected forces have been found. For instance, the strong dependence of the type of divalent ion on the strength of bridging attraction between phosphates (see for example Chap. 2, this volume). In addition to the non specific interactions there are also the specific chemical interactions which occur between ligands and receptors and proteins (Chap. 3, this volume).

In the context of cell adhesion the surfaces of interest are the surfaces of cells in an aqueous medium. This is an enormously complex surface, and considering this, it is impossible to predict the strength and range of surface forces in all but the simplest systems, the prospect of gaining understanding of the interactions between live cells and solid substrates, or cell-cell interactions appears daunting. The direct measurement of forces between two model surfaces provides one of the only ways to characterise the types of interactions which exist between surfaces and how they depend on the medium. It has been possible to study a wide range of important biological molecules such as phospholipids (Marra and Israelachvili 1985), proteins (Claesson et al. 1989) and glycolipids (Marra 1986) including glycosphingolipids (Parker 1990).

The first apparatus to measure surface forces was constructed in the 1950's by Derjaguin (Derjaguin et al. 1978) and co-workers. Surface force measurement was not made routine until the instrument of Israelachvili (Israelachvili and Adams 1978) and later variants were developed (Parker et al. 1989). More recently, the Atomic Force Microscope (AFM) has developed to the point where there are a number of commercial suppliers of instruments. The major difference between the AFM and SFA is the replacement of a surface with an atomically sharp tip and three dimensional control of the tip position. Recently the atomic force microscope has been used to measure the force between a colloid probe and a planar surface (Ducker et al. 1991).

This chapter is concerned with the techniques used to acquire surface force data. The magnitude of the forces and their range places stringent demands on the instrumentation used to measure them. A surface force apparatus (SFA) consists of four essential components :

1.   a method to control the separation between the surfaces
2.   a technique for determining the force

3.   a method for measuring the surface separation
4.   last and perhaps most important are the surfaces themselves

As many of the interactions have a molecular origin, the separation control and measurements need to have a resolution in the Ångstrom level. Curved surfaces are normally used simply because this eliminates the need to align plates parallel to within an Ångstrom and to avoid edge effects. As a result the force resolution required depends to a large extent on the radius of curvature of the surfaces used. For a 2 cm radius of curvature then 10^{-7} N force resolution is adequate to measure most forces but if the radius of curvature is reduced then higher resolutions are required.

5.2 Separation Control

There are quite a number of different ways in which the motion of a surface or tip can be controlled with very high precision. Figure 1 shows diagrammatically the common techniques used in surface force measurement. Each of the various techniques has its own advantages and disadvantages. Perhaps the most common positioning technique employs piezo electric materials. These ceramics respond to a changing electric field by changing shape. The most commonly used devices are tubes of the ceramic which have an electrode deposited on the inner surface and another deposited on the outer. Normally a high voltage supply is connected to the electrodes and the applied voltage is used to set the extension or contraction of the tube. These devices are solid state, free from vibration, and have a high frequency response (i.e. they can change their displacement very rapidly with applied voltage). Another significant advantage is that piezo tubes can also be used to achieve three dimensional positioning (Binnig and Smith 1986). Unfortunately the extension is not linear with applied voltage. This is a simple result of the fact that as the shape of the tube changes then so does its capacitance, and as a result the charge on the electrodes and the electric field. Another problem with these devices is the fact that the displacement creeps with time and some attempts have been made to eliminate these problems (Kaizuka and Byron 1988 ; Barrett and Quate 1991).

In order to avoid some of the difficulties associated with piezo ceramics, a number of other motion control schemes have been developed.

The differential spring which is employed in the Surface force apparatus developed by Israelachvili (Israelachvili and Adams 1978) provides

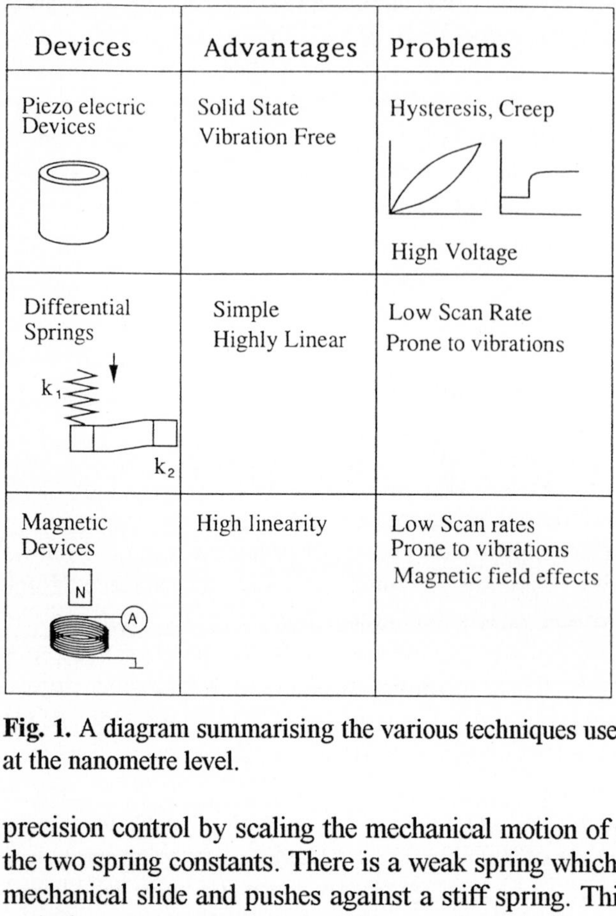

Devices	Advantages	Problems
Piezo electric Devices	Solid State Vibration Free	Hysteresis, Creep High Voltage
Differential Springs	Simple Highly Linear	Low Scan Rate Prone to vibrations
Magnetic Devices	High linearity	Low Scan rates Prone to vibrations Magnetic field effects

Fig. 1. A diagram summarising the various techniques used for position control at the nanometre level.

precision control by scaling the mechanical motion of slide by the ratio of the two spring constants. There is a weak spring which is compressed by a mechanical slide and pushes against a stiff spring. This is simple, reliable and cheap to implement, but the speed of the device is limited by the mechanical positioning of the weak spring and because of its mechanical nature, it is prone to vibrations.

Various magnetic devices have also been constructed and used. The device of Derjaguin (Derjaguin et al. 1978) employed magnetic motion control and more recently magnetic surface separation control has been implemented in a MkIV surface force apparatus (Stewart and Christenson 1990). These devices are highly linear with applied magnetic field. However, they are prone to vibrations (as the magnet is usually mounted either on a weak torsion spring or on a double cantilever spring) and they generally have a relatively low bandwidth due to induction. Precision current regulation is required to operate these devices and care has be taken so that the heating of the coils does not introduce thermal drift. Finally, the effect of strong magnetic fields on the surface forces themselves is unclear.

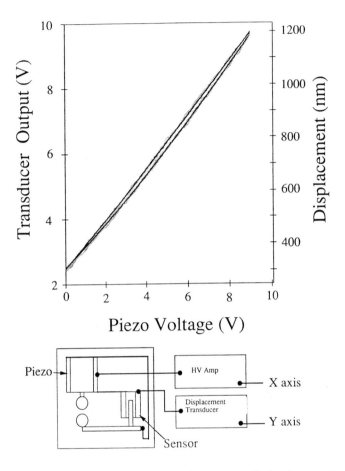

Fig. 2. The expansion of a piezo tube is plotted against the applied voltage and the response from an LVDT displacement sensor.

The best method for achieving motion control is with piezo electric devices, the advantages of these devices far outweigh the disadvantages. For accurate m̃easurements the hysteresis must be accounted for. Commercial AFMs include calibrations of the piezo hysteresis which are based on polynomial fits to the voltage displacement curve. Another way to accurately cope with hysteresis is to measure the voltage displacement curve for each measurement. This can be achieved with a number of different sensors including optical and electromechanical sensors. The resolution required to characterise the voltage displacement transfer function is not as high as required for displacement measurement because piezo devices are very linear over short extensions. One suitable and inexpensive method for measuring the displacement is to employ an LVDT (Linear Variable Differential Transformer). With the appropriate electronics it is possible to measure the hysteresis loop with reasonable precision and then a second order polynomial fit to this data can provided an accurate displacement given the voltage applied to the tube. Figure 2 shows an example of the data obtained in this way.

5.3 Force Measurement

All the techniques so far used to measure the force between two surfaces (or a tip and a surface) rely on one of the surfaces being connected to a spring. The deflection of the spring is then measured by some means and with the spring constant the force follows simply from Hookes law. Figure 3 shows schematically some of the schemes which have been implemented. For atomic force microscopy three techniques have been employed, tunnelling, laser deflection, and interferometry. With tunnelling the high sensitivity of the tunnelling current between a second tip and the cantilever is used to measure the cantilever position (for example see Kirk et al. 1988). Often the force exerted by the tunnelling tip can exceed the surface force of interest and this complicates the interpretation of data produced by such devices. The laser deflection technique is very commonly used (Meyer and Amer 1988). It is relatively easy and cheap to implement ; however, very short cantilevers are required for maximum sensitivity. For surface force measurements with the SFA by far the best method for determining the deflection of the cantilever and the separation between the surfaces is fringes of equal chromatic order, FECO, interferometry (Israelachvili 1973). This technique has the advantage that the separation between the surfaces is determined directly along with the surface shape and it is also possible to determine the refractive index of the liquid between the surfaces. The only drawback with this technique is that it requires thin transparent sheets to perform the measurements, as the interferometer is formed by silvering the back of the sheets. Also the radius of curvature of the surfaces needs to be reasonably large in order to see the fringe pattern and as a result the measurements need to be performed slowly.

Devices	Advantages	Problems
Tunnelling 	Solid State Small size.	Prone to surface Contaminants. Often Tunnelling tip AFM tip interaction > AFM tip surface interaction
FECO 	Direct Refractive Index Surface Shape	Thin (2-3μm) Transparent Sheets Required Not well suited to highly reflective surfaces
Laser Deflection 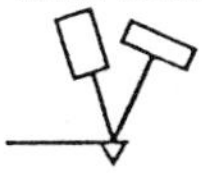	Simple	Works best with very short cantilevers.
Interferometers	Very high resolution	Complicated Set up.
Bimorphs	Simple	Single Cantilever High impedance charge measurement for DC operation.

Fig. 3. Various displacement and force detection schemes employed in surface force and atomic force microscopy measurements are shown diagrammatically.

Another technique which has been used to perform surface force measurements is to employ a piezo electric bimorph (Parker 1992). The deformation of a piezoelectric material produces an electric field in a manner converse to the shape change in response to an applied electric field. The charge developed can be used to measure displacement and force, and the sensitivity can be increased by employing two slabs of piezo material fastened together with their polarisation directions facing each other. If one end of a piezo bimorph is deflected by applying a force then a compressive strain is produced in one slab and an expansion is produced in the other and a charge develops. Bimorphs were first used by Tabor and Israelachvili to measure van der Waals interactions between two mica surfaces using a dynamic method, and they have also been used for dynamic measurements in polymer melts (Israelachvili al. 1989). With the advent of high grade electrometer amplifiers it has only recently become possible to use a bimorph to measure static surface forces (see Parker 1992, for a description of the technique). The bimorph has the advantage that the device is very simple, solid state, very cheap to implement and can be used under almost any conditions. Because it is not an optical device it can be used in any fluid and with any type of surface.

5.4 Description of the Surface Force Apparatus

A schematic diagram of a surface force apparatus based on the bimorph measurement technique is shown in Fig. 4. One surface is mounted at the end of the sensor and the other is mounted at the end of a piezo electric tube. The bimorph is enclosed in a teflon sheath and this is mounted inside a small chamber (volume ~ 10 ml). The chamber is clamped to a translation stage which is used to control the coarse position of the piezo electric tube and the upper surface. The apparatus has a foot print of 6.5 by 5 cm and is mounted inside an aluminium enclosure for temperature regulation.

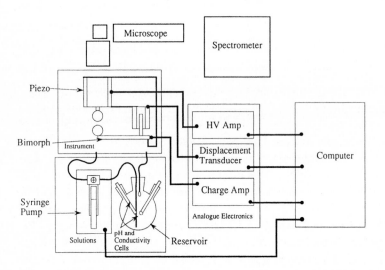

Fig. 4. A block diagram of the surface force apparatus. One surface is mounted at the end of a piezo electric bimorph force sensor the other is mounted at the end of a piezo electric tube. A linear displacement sensor is used to calibrate the response of the piezo tube to a voltage applied from the high voltage amplifier. A personal computer is used to control the entire system. It outputs a wave form to a high voltage amplifier and records the piezo displacement and bimorph charge in response to this. A syringe pump is used to recirculate solution from a reservoir into the force measuring chamber.

A computer generated waveform is used to control a high voltage amplifier, which in turn drives a piezo electric tube. The signal from the bimorph charge amplifier, and the piezo motion are recorded in response to the voltage applied to the piezo tube. With this system the speed of motion of the upper surface and the number of points sampled can be varied over a wide range from fractions of seconds to hours. The chamber of the apparatus is designed for use with an external reservoir and a syringe pump is used to circulate the solution .

The apparatus can be used with any type of hard smooth surface. Glass is the favoured substrate for surface force measurements because the surfaces are very easy to prepare by simply melting the end of a glass rod in a gas burner until a molten droplet of glass has formed. During melting, the surface tension of the glass droplet pulls the surface smooth, and when the sample cools it solidifies and retains its smoothness. A long working distance optical microscope can be used to inspect the surfaces or to focus light into a spectrometer. With this arrangement it is possible to use mica like substrate and interferometry to measure surface separation.

5.5 Force Measurements

Surface forces are recorded in a continuous manner. An example of the data produced by the device is shown in Fig. 5. The displacement of the piezo tube is plotted in response to the voltage output from the bimorph charge amplifier (which is in turn proportional to force) in Fig. 5A. Large values on the X-scale correspond to large surface separations. The surfaces are pushed together until they come to contact. When the surfaces are in contact the motion of the piezo tube is transmitted directly to the sensor. This region of the curve can be used to calibrate the sensitivity of the sensor. It is then a simple matter to calculate the surface separation from the motion of the piezo and the deflection of the bimorph calculated from the calibration. Using this procedure it is not possible to define the separation with respect to the contact between the two surfaces in dry atmosphere. As a result it is not possible to determine the thickness of an adsorbed layer if it is not pushed away from the contact region during the experiment.

The force between the two surfaces is given by the sum of both the hydrodynamic and static contributions. The hydrodynamic force arises

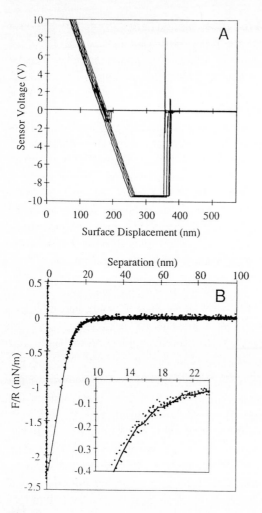

Fig. 5. An example of the raw data collected during a force measurement between two glass surfaces in $5x10^{-5}$M CTAB solution. The bimorph response is plotted against the calibrated displacement of the piezo tube in **Fig. 5A**. At large surface separations there is no surface force present and the bimorph charge remains constant. When a force is encountered the bimorph begins to deflect and a response is measured. When the two surfaces come into contact the upper surface, i.e. the surface mounted at the end of the piezo, pushes the lower surface at a constant rate. Four separate measurements are recorded in quick succession, all are displayed. The slope of the line when the surfaces are in contact remains constant. They are slightly displaced from each other due to a small thermal drift between the measurements. The forces calculated from the raw data are shown in **Fig. 5B**. The data points are measured from all four force runs and the *solid line* is the average of the data.

because of the drainage of liquid from between the surfaces (Chan and Horn 1985) and is for crossed cylinders given by:

$$F_H = \frac{6\pi R_g R_h \eta}{D(t)} \frac{dD}{dt}$$ (1)

where R_g and R_h are the geometric and harmonic radius of curvature of the surfaces and η the viscosity of the liquid. As the separation $D(t)$ becomes small the hydrodynamic force increases dramatically. This ultimately restricts the speed at which surface force measurements can be made. The simplest way to reduce the hydrodynamic force and thus increase the measuring speed is to reduce the radius of curvature of the surfaces. A reduction of radius by one order of magnitude will reduce the hydrodynamic force by two orders of magnitude, whereas the static surface force is reduced by only one order of magnitude. With surfaces with a radius of 1 to 2 mm it is possible to measure static equilibrium forces 10 times faster than is possible with a radius of 2 cm. Not only can the measurements themselves be performed more quickly, but also the time between successive runs can be very much reduced. This has tremendous benefits including lower sensitivity to thermal drift, and it allows the collection and averaging of data from many separate force measurements and in this way the signal to noise ratio can be improved.

There is an enormous amount of new information which can be obtained with higher speed surface force measurements. For instance, higher speed measurements could potentially allow the dynamics of surface interactions to be measured and rearrangements in thin films to be measured and quantified.

5.5.1 Interactions Between Silica Surfaces in Surfactant Solutions

When silica is immersed in water, the surface becomes negatively charged due to the dissociation of silanol groups. Addition of a cationic surfactant CTAB (Cetyl trimethyl ammonium bromide) reduces the surface charge through adsorption on to the negative sites. The adsorption process can be monitored by measuring the forces between the surfaces (Parker and Claesson 1994). Addition of CTAB at low concentrations has two effects on the interaction between the surfaces, a lowering of surface charge leads to a reduced repulsive double layer force, and the attraction between the surfaces at small separations becomes larger. The surfactant molecules adsorb with the headgroups on the surface and the tails of the molecules pointing out toward the aqueous solution. With the molecules adsorbed in this orientation the surfaces become hydrophobic. At a concentration of $5x10^{-5}M$ (pH=6), the surface charge on glass is neutralised and an attractive force in excess of that expected from van der Waals interactions (results shown in Fig. 5) is measured. The origin of the attraction between two hydrophobic surfaces remains unclear although several explanations have been suggested. The mechanism is still the subject of much debate more than 10 years after the discovery of long range attractive forces between hydrophobic surfaces in water.

Some of the suggestions for the origin of the hydrophobic interaction are listed below:

1. Ordering of water between two hydrophobic surfaces (Eriksson et al, 1989).
2. Electrostatic correlations between dipoles originating either in the water close to the hydrophobic surface or in patches of oriented adsorbed molecules.(Attard 1989 ; Tsao et al, 1993).
3. It has also been suggested that the interaction is linked to the formation of a vapour cavity (Christenson and Claesson 1988 ; Berard et al 1993) which occurs when two hydrophobic surfaces are brought into contact.
4. Correlations in the fluctuations of the water interface formed at a thin vapour layer separating the surface from the medium. (Ruckenstein and Churaev 1991)

A great deal of experimental effort has been directed at testing the various proposed ideas. For instance, an electrostatic correlation mechanism can be dismissed on the basis of the behaviour of the force in added electrolyte (Christenson et al. 1990 ; Parker and Cleasson 1993). Other proposed mechanisms are more difficult to test (for example 1 and 3) but with the advent of new surface force measurement techniques more experimental data will become available.

Returning now to the forces measured in CTAB solutions, a further increase in the surfactant concentration leads to further adsorption of surfactant (Parker et al 1993 ; Rutland and Parker 1993). After the neutral point is reached adsorption occurs not through electrostatic interactions but through hydrophobic interactions between the tails of the adsorbing molecules and the surface. At concentrations just below the critical micelle concentration ($\sim 1x10^{-3}$ M), CTAB adsorbs to the hydrophobic interface to form loosely packed bilayers. The surfaces consist of one adsorbed monolayer with surfactant head groups pointed toward the surface and on this a second layer with the surfactants oriented

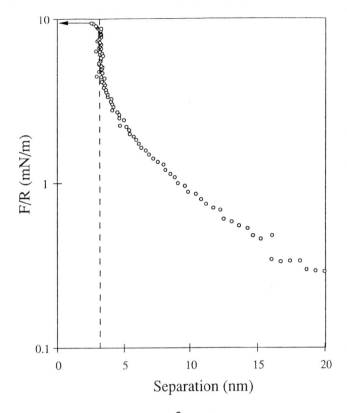

Fig. 6. Forces between two glass surfaces at 1×10^{-3} M CTAB solutions. For a full description of the results see Rutland and Parker (1993).

with the head groups pointing toward the aqueous phase. The forces measured between these surfaces are exponentially repulsive and can be well fitted with DLVO theory (results shown in Fig. 6) down to small separations. At a separation of ~3 nm a small steric barrier is encountered and the force curve turns sharply upward. This is the position where the surfactant bilayers first start to contact. At a certain applied force the surfactant monolayer yields and the surfaces move into adhesive contact. This is due to the compression and removal of the second layer of the bilayer. The hemi-fusion of supported bilayers have been extensively studied with surface force techniques (Horn 1984 ; Helm et al. 1989).

References

Attard P (1989) The hydrophobic interaction. J Phys Chem 93: 6441-6444

Barrett RC, Quate CF (1991) Optical scan-correction system applied to atomic force microscopy. Rev Sci Instrum 62:1393-1399

Berard DR, Attard P, Patey GN (1993) Cavitation of a Lennard-Jones fluid between hard walls, and the possible relevance to the attraction measured between hydrophobic surfaces. J Chem Phys 98: 7236-7241

Binnig, G, Smith,D (1986) Single-tube three-dimensional scanner for scanning tunneling microscopy. Rev Sci Instrum 57: 1688-1689

Chan DYC, Horn RG (1985) The drainage of thin liquid films between solid surfaces. J Chem Phys 83: 5311-5327

Christenson HK., Claesson PM (1988) Cavitation and the interaction between macroscopic hydrophobic surfaces. Science 239: 390-392

Christenson HK., Fang J, Ninham BW, Parker JL (1990) Effect of divalent electrolyte on the hydrophobic attraction. J Phys Chem 94: 8004-8006

Claesson PM, Blom CE, Herder PC, Ninham BW (1986) Interactions between water-stable hydrophobic Langmuir-Blodgett monolayers on mica. J Colloid Interface Sci 114: 234-242

Claesson PM, Arnebrant T, Bergenståhl B, Nylander T (1989) Direct measurements of the interaction between layers of insulin adsorbed on hydrophobic surfaces. J Colloid Interface Sci 130: 457-466

Derjaguin BV, Rabinovich YI, Churaev NV (1978) Direct measurement of molecular forces. Nature 272: 313-318

Ducker W, Senden T, Pashley RM (1991) Direct measurement of colloidal forces using an atomic force microscope. Nature 353: 239-241

Eriksson, JC, Ljunggren S, Claesson PM (1989) A phenomenological theory of long-range hydrophobic attraction forces based on a square-gradient variational approach. J Chem Soc Faraday Trans II. 85: 163-176

Helm CA, Israelachvili JN, McGuiggan PM (1989) Molecular mechanism and forces involved in the adhesion and fusion of amphiphilic bilayers. Science 246: 919-922

Horn, RG (1984) Direct measurement of the force between two lipid bilayers and observation of their fusion. Biochim Biophys Acta. 778: 224-228

Israelachvili JN (1973) Thin film studies using multiple-beam interferometry." J Colloid Interface Sci 44: 259-272

Israelachvili JN (1991) Intermolecular and surface forces. 2nd ed. Academic Press, London

Israelachvili JN, Adams GE (1978) Measurement of forces between two mica surfaces in aqueos electrolyte solutions in the range 0-100 nm. J Chem Soc Farad Trans 1 74: 975-1001

Israelachvili JN, S. J. Kott SJ, Fetters LJ (1989) Measurement of dynamic interactions in thin films of polymer melts: the transition from simple to complex behavior. J Polymer Sci, Part B 27: 489-502

Kaizuka HB, Byron S (1988) A simple way to reduce hysteresis and creep when using piezo electric actuators. Jap J Applied Phys 27: L773-L776

Kirk MD, Albrecht TR, Quate CF (1988) Low-temperature atomic force microscopy. Rev Sci Instrum 59: 833-835

Marra J (1986) Direct measurement of the interaction between phosphatidyl-glycerol bilayers in aqueous electrolyte solutions. Biophys J 50: 815-825

Marra J, Israelachvili J (1985) Direct measurement of forces between phosphatidylcholine and phosphatidylethanolamine bilayers in aqueous electrolyte solutions. Biochemistry 24: 4608-4618

Meyer G, Amer N (1988) Novel optical approach to atomic force microscopy. Appl Phys Lett 53: 1045-1047

Parker JL (1990). Forces between bilayers containing Charged glycolipids. J Colloid Interface Sci 137: 571-576

Parker JL (1992) A novel method for measuring the force between two surfaces in a surface force apparatus. Langmuir 8: 551-556

Parker JL, Claesson PM (1994) Forces between hydrophobic silanated glass surfaces. Langmuir 10:635-639

Parker JL, Christenson HK, Ninham BW (1989) Device for measuring the force and separation between two surfaces down to molecular separations. Rev Sci Instrum 60: 3135-3138

Parker JL, Yaminsky V, Claesson PM (1993) Surface Forces Between Glass Surfaces in Cetyltrimethylammonium Bromide Solutions. J Phys Chem 97: 7706-7710

Ruckenstein E, Churaev E (1991) A possible hydrodynamic origin of the forces of hydrophobic attraction. J Colloid Interface Sci : 535-538

Rutland MW, Parker J (1994) Surface forces between silica surfaces in cationic surfactant solutions: bilayer formation and effect of pH. Langmuir in press

Stewart AM, Christenson HK (1990) Use of magnetic forces to control distance in a surface force apparatus. Meas Sci Technol 1: 1301-1303

Tsao YH, Evans DF, Wennerström H (1993) Long-Range attraction between a hydrophobic surface and a polar surface is stronger than that between two hydrophobic surfaces. Langmuir 9: 779-785

6 Measurement of the Strength of Cell-Cell and Cell-Substratum Adhesion with Simple Methods

A. M. Benoliel, C. Capo, J. L. Mège and P. Bongrand

6.1 Introduction

There is an obvious need for scientists to use a clear definition of the phenomena they study. Thus, a cell may be considered as bound to another cell only by showing that these cannot be separated by a force of intensity F applied during a given time t. Since there is no absolute definition of adhesion, discordant results might be obtained by investigators exploring the same cellular system with different experimental criteria for attachment. Therefore, it is useful to describe accessible methods allowing quantitative or semi-quantitative determination of binding strength. We shall describe a set of techniques that have been used in our laboratory for more than 10 years. The basic idea is to generate calibrated hydrodynamic flows subjecting cells to mechanical forces that can be evaluated numerically. The procedures we shall describe require no specialized equipment and may be easily performed in any standard laboratory. However, only semi-quantitative estimates of the forces experienced by cells can be obtained. More refined approaches are described by Goldsmith and Amblard (chapters 10 and 7). However, they are based on sophisticated custom-made equipment.

6.2 Determination of the Strength of Cell Aggregates

The basic principle of the method we shall now describe consists of subjecting cell aggregates to strong shear in a thin syringe needle. This technique was successfully applied to the study of interactions between phagocytic cells and diverse particles (Capo et al. 1978 ; Bongrand et al. 1979), conjugates made between cytotoxic lymphocytes and target cells (Bongrand and Golstein 1983 ; Bongrand et al. 1983), concanavalin A-agglutinated thymocytes (Capo et al. 1982) or tumor cells (Capo et al. 1985). We shall first recall fundamental results from fluid mechanics. Practical methods for generation of calibrated flows and selected examples will then be described.

6.2.1 Laminar Flow in a Cylindrical Pipe

When a fluid is driven into a cylindrical pipe of radius a with low enough pressure (see below for more details concerning stability), the fluid velocity is everywhere parallel to the pipe axis and its magnitude v at any point M is given by the well-known Hagen Poiseuille formula (see e.g. Sommerfeld 1964) :

$$v = 2Q(a^2 - r^2)/\pi a^4 \tag{1}$$

where r is the distance between M and the pipe axis and Q is the flow rate (in m³/second). The velocity gradient or shear rate G=dv/dr is zero on the

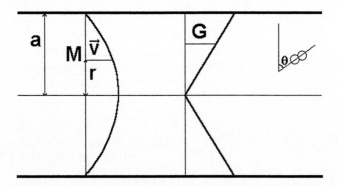

Fig. 1. Hydrodynamic flow in a cylindrical pipe of radius **a**. The velocity profile is parabolic with zero velocity near the wall. The shear rate **G** is maximum near the wall and zero on the axis. **V** is the velocity at point **M**, at distance **r** from the axis.

pipe axis and it is maximum near the wall. Another useful formula is the relationship between the flow rate Q, the driving pressure P and pipe length L :

$$Q = P \, \pi a^4 / 8 \mu L \tag{2}$$

where μ is the medium viscosity (in pascal.second). Thus, the maximal shear rate obtained by applying pressure P is :

$$G_{max} = Pa/2\mu L \tag{3}$$

Now, let us consider a doublet of cells modeled as identical spheres of radius b (Fig. 1) exposed to the flow. If b is much smaller than a, the shear rate G may be considered as constant near the doublet. It may be shown that cells may experience a separative force depending on the angle θ between the doublet axis and the fluid velocity gradient (Arp and Mason 1977 ; Bongrand et al. 1988 ; Goldsmith Chap. 10). Although somewhat intricate calculations are needed to derive this force, an order of magnitude may be obtained as follows : by reason of symmetry, the doublet velocity is expected to be equal to the velocity of undisturbed fluid flow on the midpoint between the sphere centers. Neglecting the fluid disturbance by the doublet, each sphere is expected to experience a hydrodynamic drag proportional to the relative flow velocity $Gb\sin\theta$ on the sphere center. Tentatively using Stokes' law, the force may be approximated as $6\pi\mu Gb^2\sin\theta$, and its component parallel to the doublet axis would be $6\pi\mu Gb^2\sin\theta\cos\theta$. The maximum distractive force would thus be equal to 9.4 Gb2 (corresponding to an angle θ of $\pi/4$). This is quite similar to the "exact" value of 19.2 μGb^2. Note that the doublet is also subjected to a tangential disruptive force (i.e. perpendicular to the doublet axis) of similar order of magnitude. This is neglected in the present semi-quantitative estimate. The basis of our method is to drive cell aggregates through a syringe needle with fairly high and controlled pressure and measure the fraction of released elements. Data interpretation raises two important problems.

First, experimental studies showed that the equations we described are no longer valid when the flow velocity is too high, due to the generation of turbulences. An empirical way of predicting this phenomenon is to calculate Reynold's number R:

$$R = v \, a \, \rho/\mu \tag{4}$$

where v and a are a velocity and a linear dimension characteristic of the system, ρ and μ are the volumic mass and viscosity of the medium. When the pipe inlet is fairly regular, turbulences do not occur when R is lower than about 20,000 (Sommerfeld 1964). Using equations (1) and (3), The corresponding wall shear rate is about :

$$G = 2 \ v/a = 2 \ R\mu/\rho a^2 \tag{5}$$

Thus, the maximum shear rate that can be obtained in an aqueous medium of volumic mass 1,000 kg/m3 and viscosity 0.001 Pa second such as an aqueous solution is of the order of $0.04/a^2$, where a is in meter.

A second point of importance is that the distractive force experienced by a cell doublet is dependent on its location in the pipe as well as its orientation (equation 1). What can be calculated is only the maximum force. Hence, in order to ensure that most cell aggregates were actually exposed to such a force, it was felt reasonable to drive cell suspensions several times through a needle.

A third point is that the maximum disruptive force is exerted only during a very short period of time, of the order of 1/G (Arp and Mason 1977) due to the doublet rotation. Hence, it may be expected that doublet rupture probably involves the simultaneous disruption of adhesive bonds, in contrast to slow detachment (Capo et al. 1982).

Units and constants
Two systems of units are presently used. We recall basic conversion factors :

Parameter	c.g.s. unit		S.I. unit
Viscosity	1 poise	equals	0.1 pascal.second
Force	1 dyne	equals	10^{-5} newton
Energy	1 erg	equals	10^{-7} joule

The viscosity of water is dependent on the temperature following (Weast and Astle 1982) :

Temperature (°C)	5	10	15	20	25	30	37
Viscosity (mPa.s)	1.52	1.31	1.14	1.00	0.89	0.80	0.69

6.2.2 Disruption of Cell-Cell Aggregates in a Syringe Needle

Generation of a calibrated pressure
A very simple way of generating calibrated flows is to exert a constant pressure on the piston of a syringe needle maintained vertical on a suitable holder (Capo et al. 1978). This was performed by depositing on the syringe piston an Erlenmeyer flask (100 to 400 ml) containing various amounts of water. The pressure P is simply :

$$P = mg/A \tag{6}$$

where m is the total mass of the piston and deposited weight, g is 9.81 m/s^2 and A is the piston area.

When aggregates are very weak, the pressure generated by this procedure is too high. A very convenient alternative procedure consists of using automatic pipettes. As shown in Fig. 2, an automatic pipette has a "dead volume" V_0 (once the removable cone is set). To aspirate a volume V_1 of water, the piston is pushed and released after dipping the cone into watter. Following Boyle-Mariotte's law, the aspiration pressure is then :

$$P = P_0 \, V_1/V_0 \tag{7}$$

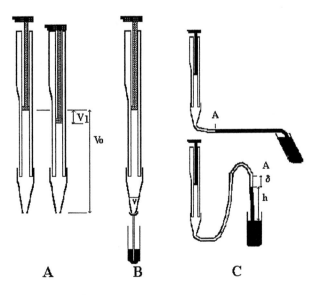

Fig. 2. Calibration of an automatic pipette.

where P_0 is the atmospheric pressure (i.e. about 10^5 Pa). The maximum pressure generated by the pipette can thus be determined by measuring V_0. We suggest two ways of achieving this goal :

1st method. A calibrated needle is set on a cone that is fixed on the pipette (Fig. 2B). The piston is pushed with setting V_1, then released when dipping in a tube containing a volume $V_1/2$ of water. The time t required for complete aspiration is measured with a stopwatch. It may be shown (Bongrand and Golstein 1983) that :

$$V_0 = V_1 + (\pi a^4 P_0 \, t/8\mu L - V_1/2)/\ln 2 \qquad (8)$$

2nd method. The pipette is connected to a water-filled bucket through a long flexible tube of small section area s. A convenient amount of water is aspirated and the position of the water boundary is recorded for two configurations of the tube (Fig. 2C). When the tube is maintained vertical, the water boundary recedes by δ. We may write, neglecting the volume of air within the tube :

$$V_0 = P_0 \, \delta \, s/\rho \, g \, h \qquad (9)$$

Calibration of a syringe needle

A simple way of determining the inner diameter of a needle by measuring a flow with a liquid of known viscosity was described in a previous report (Bongrand and Golstein 1983). This is made still simpler as follows. Take a syringe of about 30 ml capacity. Put it vertical on a holder (after removing the piston). Set the needle dipping in water and fill the syringe with water. The flow rate Q is determined with a stopwatch, by looking at the time of passage of the water boundary along the syringe graduations. The needle length L is easily measured. The height h of the water level (within the syringe) as compared to the needle tip is easily measured. We may the calculate the needle inside radius a with the following formula :

$$Q = \rho g h \, \pi r^4 / 8\mu L \qquad (10)$$

this is indeed equivalent to equation (2).

An alternative way is to examine an enlarged image of the needle. We achieved this by using a copier with enlarging capacity. A small syringe was xeroxed (the only trick is to put the needle below a dark surface) Sequential enlargements are done until the diameter can be measured with a plain ruler. For convenience, a ruler with millimeter scale is copied at the same time.

Typical results

As shown in Fig. 3, it was possible to achieve graded disruption of cell-cell aggregates by subjecting them to flows of increasing velocity. Maximum shear rates on the order of 100,000 s^{-1} or more could be obtained. Syringes of 1 ml or 2 ml capacity were used in most cases. Glass syringes must be used, since the friction of plastic syringes prevents any accurate determination of the intensity of the force generating the piston displacement. Needles used for standard intravenous or intradermal injections were found convenient (10-40 mm length, 0.3-0.6 mm inside diameter).

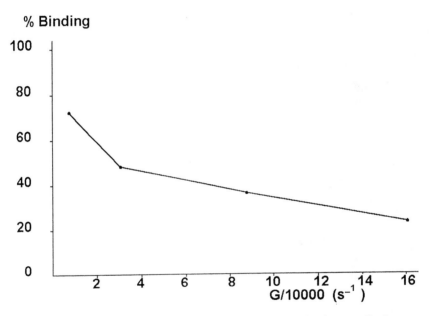

Fig. 3. Dissociation of cell aggregates by hydrodynamic forces. Conjugates made between cytotoxic lymphocytes and target cells (Bongrand and Golstein 1983) were driven through a syringe needle with varying pressure. The percent of target cells remaining bound was plotted versus maximum shear rate in the needle.

- Due to the approximate validity of Reynold's criterion it may be felt useful to check that the flow is laminar by showing that the flow rate is proportional to the weight deposited on the piston (Bongrand et al. 1979).
- In our hands, almost maximum disruption of aggregates occurred before the third passage through the syringe. Therefore, it does not seem

warranted to perform numerous passages.

- In some cases, very low flow rates are required. These can be obtained by mounting a needle on the tip of an automatic pipette as previously described (Bongrand and Golstein 1983). In fact, this procedure was found most useful to perform reproducible disssociation of large aggregates in cell pellets after centrifugation.

- The limiting parameter of aggregate strength is often the mechanical resistance of the cell membrane. This was repeatedly checked by labeling cells with radioactive chromium before shearing. Aggregate disruption often results in substantial release of cytoplasmic radioactivity when shear rates are higher than about 100,000 second^{-1}.

6.3 - Determining the Strength of Cell-Substratum Adhesion

The strength of cell-substrate adhesion was studied by subjecting cells to moderate hydrodynamic forces for fairly long periods of time in custom-made flow chambers (Weiss 1961 ; van Kooten et al. 1992). We shall briefly describe a methodology that was used to achieve rapid detachment of bound cells with strong hydrodynamic forces. This can be performed with standard material available in every cell biology laboratories.

6.3.1 Theoretical Background

We shall first present a rough estimate of the force exerted on a cell bound to the wall of a cylindrical pipe. Two extreme cases may be considered (Fig. 4).

- If the cell does not exhibit substantial spreading, it may be modeled as a sphere of radius b. The hydrodynamic drag generated by a flow of wall shear rate G is (see e.g. Goldman et al. 1967) :

$$F = 32 \mu b^2 G \tag{11}$$

where μ is the medium viscosity. Now, cell adhesion is often followed by spreading. A perfectly flat cell of area A would be subjected to a force of intensity $G\mu A$. A reasonable order of magnitude for A may be obtained by modeling the spread cell as a disk of radius 2b (In a recent study, melanoma cells exhibited about twofold apparent radius increase after deposition on a fibronectin-coated surface : André et al. 1990). Replacing A with $4\pi b^2$, we obtain :

$$F = 12.6 \, \mu \, b^2 \, G \tag{12}$$

Formulae (11) and (12) yield a similar order of magnitude for the disruptive force.

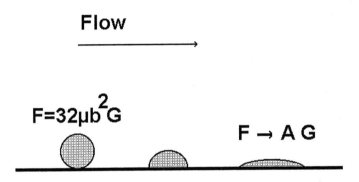

Fig. 4. Hydrodynamic force experienced by a substratum-bound cell exposed to an hydrodynamic flow.

It may be of interest to estimate the maximum force that can be obtained to detach bound cells from a capillary tube of radius a, in order to comply with Reynolds' criterion. Using equation (4) and taking as 20,000 the maximum value of Reynold's number, we obtain :

$$F_{max} \text{ (spread cells)} = 5.10^5 \, (\mu b/a)^2 \, /\rho \tag{13}$$

In a capillary tube of 1 mm diameter, the maximum force exerted on a cell of 5 μm radius by a fluid of 1000 kg/m^3 volumic mass and 0.001 Pa.second viscosity (such as water) is 12.5 nanonewton. This may be too low to detach cells, and it was found useful to increase this value by using higly viscous fluids. Thus, the above limit could be increased by more than 100 when high molecular weight dextran (500,000 MW, Pharmacia, Uppsala, Sweden) was added at about 10 % w/w (Mège et al. 1986). Note that osmotic effects may change cell-cell interactions and artefacts must be ruled out (We are grateful to Dr. Evan Evans for pointing out this possibility).

6.3.2 Examples

The method we described was applied to measure the binding strength of macrophage-like cell lines (Mège et al. 1986), tumor cells (Leung-Tak et al. 1988 ; André et al. 1990) or blood granulocytes (Mège et al. 1990) to bare or protein-coated surfaces. A standard procedure was as follows : we made use of glass capillary tubes usually employed to perform hematocrit counts. Length is about 10 cm and inside diameter is about 1 mm. Tubes were first incubated with suitable solutions of adhesion molecules or albumin. Indeed, since spontaneous adsorption occurs within a few seconds when a clean surface is exposed to biological material, it is important to work with fairly well defined coatings. Cell suspensions (about one million/ml) were then added and tubes were incubated for several tens of minutes at 37°C in a moist atmosphere. Since the tube diameter is quite small, cell sedimentation is completed within a few minutes and adhesion proceeds rapidly.

Detachment may be performed by driving a warm dextran solution through the capillary tubes with a syringe, as described above. The simplest way of quantifying adhesion is to stain cell with standard reagents (Diff Quick, Merz Dade, Duedingen, Switzerland, proved quite satisfactory) Tubes can then be observed with a standard microscope under low magnification (10 X objective), using a micrometer eyepiece to

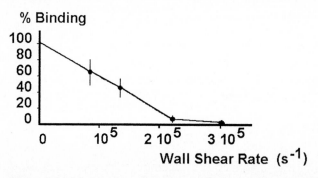

Fig. 5. Rupture of cell-substratum adhesion. Murine macrophage-like P388D1 cells were made to adhere to glass capillary tubes and subjected to hydrodynamic forces. The percentage of cells remaining bound is plotted versus wall shear stress (i. e. shear rate multiplied by fluid viscosity)

count the number of bound cells per unit length. Alternatively, cells may be labeled with radioactive chromium (incubate for 1-2 hours in 25 μCi sodium chromate per ml and wash extensively). Bound radioactivity can then be determined with a gamma counter. The latter procedure allows simultaneous determination of cell damage by centrifuging eluates and assaying cell bound and free radioactivity. In order to interpret results safely, it is important to measure the spontaneous chromium release, since this occurs in absence of any cell damage and may display wide variations depending on the nature of studied cells. A typical detachment curve obtained with macrophage-like P388D1 cells is shown in Fig. 5.

6.4 Conclusion

We described simple methods that may be used to achieve controlled disruption of cell-cell or cell-substratum bonds with a material available in any standard biological laboratory. It is thus possible to obtain rough estimates of the absolute binding strength. It is important to bear in mind that these procedures rely on the brief application of high distractive forces, resulting in the simultaneous rupture of many bonds and/or uprooting of membrane molecules and membrane tearing. Lower forces might induce cell detachement as a consequence of progressive bond rupture and continuous structural reorganizations (see e.g. Tozeren et al., 1989). This process is much more difficult to describe with quantitative modeling.

References

André P, Capo C, Benoliel A M, Bongrand P, Rouge F, Aubert C (1990) Splitting cell adhesiveness into independent measurable parameters by comparing ten human melanoma cell lines. Cell Biophys 17:163-180

Arp P A, Mason S G (1977) The kinetics of flowing dispersions. VIII Doublets of rigid spheres (theoretical). J Colloid and Interface Sci 61:21-43

Bongrand P, Golstein P (1983) Reproducible dissociation of cellular aggregates with a wide range of calibrated shear forces : application to cytolytic lymphocyte-target cell conjugates. J Immunological Methods 58:209-224

Bongrand P, Capo C, Benoliel A M, Depieds R (1979) Evaluation of intercellular adhesion with a very simple technique. J Immunological Methods 28:133-141

Bongrand P, Pierres M, Golstein P (1983) T cell-mediated cytolysis : on the strength of effector-target cell interaction. Eur J Immunol 13:424-429

Bongrand P, Capo C, Mège J L, Benoliel A M (1988) Use of hydrodynamic flows to study cell adhesion. In Bongrand P (ed) Physical basis of cell-cell adhesion. CRC press, Boca Raton, pp 125-156

Capo C, Bongrand P, Benoliel A M, Depieds R (1978) Dependence of phagocytosis on strength of phagocyte-particle interaction. Immunology 35:177-182

Capo C, Garrouste F, Benoliel A M, Bongrand P, Ryter A, Bell G I (1982) Concanavalin A-mediated agglutination : a model for a quantitative study of cell adhesion. J Cell Sci 56:21-48

Capo C, Benoliel A M, Bongrand P, Mishal Z, Berebbi M (1985) T-cell-fibroblast hybridoma deformability and concanavalin A-induced agglutination. Immunological Investigations 14:27-40

Goldman A J, Cox R G, Brenner H (1967) Slow viscous motion of a sphere parallel to a plane wall. II. Couette flow. Chem Engn Sci 22:653-660

Leung-Tak J, Capo C, DeLapeyrière O, Benoliel A M, Arnaud D, Bongrand P (1988) Relationship between cellular adhesiveness and metastatic activity in polyomavirus-transformed FR3T3 rat cell lines. Int J Cancer 42:946-951

Mège J L, Capo C, Benoliel A M, Bongrand P (1986) Determination of binding strength and kinetics of binding initiation - a model study made on the adhesive properties of P388D1 macrophage-like cells. Cell Biophys 8:141-160

Mège J L, Capo C, André P, Benoliel A M, Bongrand P (1990) Mechanisms of leukocyte adhesion. Biorheology 27:433-444

Sommerfeld A (1964) Mechanics of deformable bodies. Academic Press. New York

Tozeren A, Sung K L P, Chien S (1989) Theoretical and experimental studies on cross bridge migration during cell disaggregation. Biophys J 55:479-487

Van Kooten T G, Schakenraad J M, van der Mei H C, Busscher J (1992) Development and use of a parallel-plate flow chamber for studying cellular adhesion to solid surfaces. J Biomed Mater Res 26:725-738

Weast RC, Astle MJ (1982) Handbook of chemistry and physics. CRC press. Boca Raton. Florida

Weiss L (1961) The measurement of cell adhesion. Exp Cell Res Suppl 8:141-153

7 Method to Assess the Strength of Cell-Cell Adhesion Using a Modified Flow Cytometer

F. Amblard

7.1 Introduction

Cell-cell adhesion embraces a series of phenomena which cover many aspects of biology. They can be best described as a conjunction of morphological, biochemical and physical features. Since the pioneering works on the characterization of the first cell surface molecules involved in cell adhesion of Dictyostelium or nerve cells (Beug 1973 ; Thiery 1977 ; Gerisch 1980), the advances of molecular biology techniques have yielded an every-day-increasing catalog of surface molecules involved in so-called adhesion pathways, i.e. receptor/ligand interactions which mediate adhesion.

In most cases, the cell/cell or cell/substratum adhesion assays carried out to determine if a receptor/ligand interaction effectively mediates adhesion, are based on the assessment of the adhesion efficiency, defined as the relative amount of "sticking" cells. In these assays, the mechanical parameters of the experiment are either poorly controlled or kept roughly constant but unknown. Since not only the efficiency but also the strength of adhesion have a clear physiological relevance, we designed a novel physical assay by which these two parameters can be simultaneously assessed for conjugates in liquid suspension[1]. In combination with the use of monoclonal antibodies directed against adhesion molecules expressed at the cell surface, it is then possible to dissect the adhesion strength into its distinct molecular components.

---

[1] "Conjugate" will refer to a pair of cells adhering to each other.

7.2 Principle of the Method

7.2.1 Brief Description and Physical Aspects of Flow Cytometers

A flow cytometer is an instrument in which biological particles are carried by a laminar converging fluid stream and individually pass as a narrow file by sensors that collect physical and chemical information from each individual cell (Van Dilla 1985 ; Melamed 1990). In commercial instruments[2], a focused light beam[3] hits the particle file in the sensing region, from which several optical sensors detect the light emitted or scattered by particles. This information is digitized and collected as a multiparametric data set.

Flow cytometry is essentially used to analyze and to quantify the heterogeneity of liquid suspensions containing animal cells, plant cells, bacteria, cell nuclei or chromosomes[4]. Therefore, one fundamental feature of flow cytometers is to analyze only one particle at a time, and therefore to mechanically disrupt any unwanted aggregate that could provide misleading information. This is achieved by injecting the sample suspension into a flow chamber where it is surrounded by a larger laminar sheath stream which is then forced through an abruptly constricting channel toward the sensing region (Kachel 1990) (Fig. 1). The converging geometry of the flow chamber reduces the cross-section to a circular or rectangular shape with typical linear dimensions between 50 and 300 μm (Fig. 1). The overall reduction ratio of the cross-section area in the commercial instruments ranges from 100 to 1500. The resulting reduction

---

[2] Most research and clinical investigators use one of the few types of commercially available instruments ; "home made" instruments with atypical characteristics are very rare. By default, general statements will concern here the existing commercial instruments.

[3] generally a laser beam which is perpendicular to the fluid stream and converges within it to a small cross-section with typical linear dimensions of 10-20 μm in the direction of the stream and 50-100 μm in the perpendicular direction.

[4] Some investigators have also used flow cytometry to study very small particles such as viruses, liposomes, subcellular organelles, large immune complexes, or larger objects like cell aggregates or multicellular organisms (Melamed 1990).

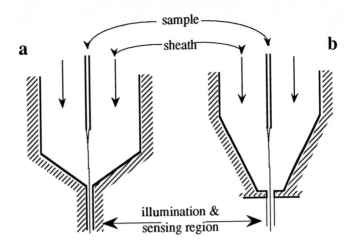

total flow rate: q
perpendicular cross-section: A(z)
fluid velocity (vector): $V(q,z,r)$
axial fluid velocity (scalar): $v(q,z,r=0)$
axial velocity gradient (scalar): $G(q,z)$

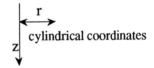

cylindrical coordinates

Fig. 1. Hydrodynamic focusing through the flow chamber. Schematic axial cross-section of two typical flow chamber geometries : a quartz flow chamber connected to a straight channel, and a nozzle which produces a cylindrical jet-in-the-air. The sample stream is injected along the axis of a laminar sheath flow which converges toward the sensing region. Typical dimensions are 150 to 300 μm for the internal diameter of the sample injection needle, 1 mm for the internal diameter of the flow chamber upstream the constriction, 50 to 300 μm for the linear size of the final cross-section of the total stream, and less than 10 to 30 μm for the diameter of the final cross-section of the sample stream. In case **b**, mean stream velocities of the jet-in-the-air are usually between 5 and 10 m.s^{-1}. In case **a**, these velocities are roughly 10 times smaller in the sensing region.

of the sample stream to a very narrow file in the center of the sheath stream is called hydrodynamic focusing : the diameter of the sample core which flows along the center of the stream is typically below 10 to 30 μm. This principle, first used by Crosland-Taylor (1953), has the essential advantage that all particles are subjected to the same set of hydrodynamic stresses, and pass through a region of uniform sensitivity of the detectors.

Optical sensing can take place either within a quartz flow chamber (Fig. 1a), or in a jet-in-the-air produced by a narrow exit orifice (Fig. 1b). Flow chambers have been designed with various geometries : either axisymmetric boundaries or rectangular cross-sections, or even an axisymmetric vertex connected to a rectangular channel (Pinkel and Stovel 1985).

Noteworthy, upstream of the flow chambers shown on Fig. 1, the sample suspension travels through a small constant diameter tubing. In most instruments and under standard analysis conditions, the shear experienced at the tubing inlet as well as through the tubing can be neglected. Indeed, typical rates of these shear stresses are roughly 3 orders of magnitude smaller than the rate of the extension undergone at the chamber constriction (Amblard 1993a).

7.2.2 Generation of Calibrated Hydrodynamic Extensions

In both cases of Fig. 1, every particle flowing along the center of the converging stream experiences a set of tremendous accelerations and high velocity gradients. The resulting extension along the axis generates viscous forces that tend to orient and to elongate particles and to disrupt aggregates. The mechanical effects of this extension will be presented below. The magnitude of this stress depends on the geometry of the flow chamber and on the flow rate q[5]. The hydrodynamic stress can be described using the gradient of the axial velocity :

$$G(z) = \frac{dv(q,z)}{dz} \tag{1}$$

where z is the abscissa on the chamber axis (Fig. 1) and $v(q,z)$ the scalar function that represents the velocity along z. If the Poiseuille approximation is valid (Batchelor 1967), it comes :

$$v(q,z) = \frac{2q}{A(z)} \tag{2}$$

---

[5] By default, flow rate will refer to the total flow rate : q = sample flow rate + sheath flow rate. The sample flow rate is 2 to 4 orders of magnitude smaller than the sheath flow rate.

$$G(q,z) = \frac{-4qb'}{\pi b^3} \tag{3}$$

where A(z) is the area of the perpendicular cross-section at position z, b(z) the radius such that $A=\pi b^2$, and b'=db/dz. In the case of abruptly constricting geometries such as the "hole" shown on Fig. 1b, the Poiseuille approximation is not valid, but the overall velocity field V(q,z,r) may be computed using the exact solution developed for the viscous flow through a circular aperture in a plane wall (Happel 1983 ; White 1991). Obviously, G(q,z) is a non constant function of z which depends on the flow rate[6] and on the shape. In other words, for any given geometry of the flow chamber, G(q,z) reaches a maximum value $G_m(q)$, and that value is almost proportional to q in the laminar limit[7]. In jet-in-the-air instruments operated at standard flow rates, $G_m(q)$ can be very high : in the 10^5-10^6 s^{-1} range (Pinkel 1985).

Noteworthy, the kind of biological information expected from most flow cytometric analyses is not affected by such strong hydrodynamic extensions which meet rather well with the requirement for disrupting aggregates and for analyzing only one particle at a time. Therefore, in most instruments, the sheath flow rate is user-adjustable on a very limited range only, or even not adjustable at all.

7.2.3 Principle of the Method

From the above description, it is obvious that the very principle of flow cytometry *hydrodynamic focusing* can be used :

1. to generate calibrated axisymmetric extensions of virtually any rate,
2. to have all sample particles experience the same stress in exactly the same conditions,
3. to assess the disruption of paired particles, and thus their adhesion strength.

---

[6] In a linear fashion within the Poiseuille aproximation.

[7] This is obvious in the Poiseuille approximation. In the non-Poiseuille but still laminar case, the effect of inertia forces is to slightly shorten the high velocity gradient region by shifting it towards the flow direction. The resulting increase of $G_m(q)/q$ is neglibible as far as the flow keeps laminar.

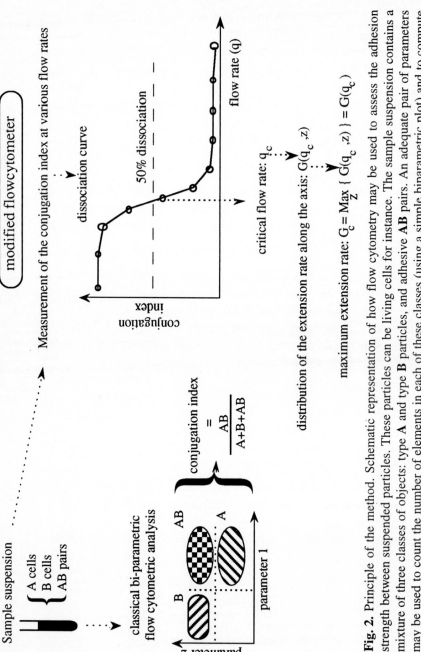

Fig. 2. Principle of the method. Schematic representation of how flow cytometry may be used to assess the adhesion strength between suspended particles. These particles can be living cells for instance. The sample suspension contains a mixture of three classes of objects: type **A** and type **B** particles, and adhesive **AB** pairs. An adequate pair of parameters may be used to count the number of elements in each of these classes (using a simple biparametric plot) and to compute a so-called conjugation index. The sample is analyzed in the modified flow cytometer at various flow rates, and the conjugation index is plotted as a function of the flow rate. From the resulting dissociation curve one extracts the critical flow rate q_c, and successively computes the axial distribution of the extension rate $G(q_c,z)$, and the maximum of this distribution with respect to **z**: $G_c = \text{Max}_z \{G(q_c,z)\}$. G_c is expressed in s^{-1} and is a quantitative indication of the

The aim of our work was to assess the adhesion strength between pairs of suspended cells. The principle of the method is illustrated on figure 2. It consists in analyzing particle suspensions under various stress conditions, and establishing so-called "dissociation" curves. From these curves, one then determines the critical extension rate beyond which cell-cell pairs are disrupted. To make this possible, we modified a commercial instrument as described below, in such a way that one could not only analyze samples at any flow rate (Amblard et al. 1992) but also precisely compute the velocity field in the whole chamber and the resulting stress imparted onto the particles by the fluid flow. One could thus extract from the dissociation curves quantitative information on the adhesion strength (Amblard 1994).

7.3 Instrumental Set-Up and Operation[8]

The flow cytometer on which the present method was developed is a FACSTAR Plus™[9]. This instrument meets with the basic requirements for the implementation of the technique:

- easy access for the modification of the flow chamber,
- direct visual control of the flow chamber and sensing region,
- stable and accurate manual pressure control over the sheath and sample streams,
- accessibility of the electronic boards to be modified,
- direct oscilloscopic control of the analog signals generated by the photodetectors,
- sophisticated and "easy to adapt" detection electronics,
- advanced circuitry for the detection of coincident signals[10].

---

[8] More detailed informations can be found in two original papers (Amblard et al. 1992)

[9] Manufactured by Becton-Dickinson, San Jose CA. It is one of the most popular instruments used in research and clinical laboratories.

[10] Particles which are too close to each other might produce partially coincident or overlapping signals. Such a source of potential artifacts must be strictly controlled.

7.3.1 Fluidic Modifications

The FACSTAR Plus™ is based on a jet-in-the-air system, with a geometry similar to that of figure 1b. The regular nozzle provided with the instrument produces very high and quasi fixed extension rates : 5.10^5 s^{-1} for the nozzle with a 70 µm exit orifice. It was replaced by a new quartz flow chamber manufactured for us by Hellma, Germany. The chamber design is presented with details in a previous publication (Amblard et al. 1992). Briefly, the new flow chamber has a geometry similar to that of a converging funnel : it is essentially made of a 30 degree revolution vertex connected to a straight channel with a 250 µm square cross-section[11]. The essential and new feature of this geometry is that the connection between the circular cross-section of the vertex and the square cross-section of the channel is very progressive. As a consequence, the whole chamber could be considered to be hydrodynamically equivalent to a virtual revolution funnel defined by a so-called equivalent radius b(z), b(z) being a "smooth" function of z^{12}. This "smooth" geometry helps to produce "soft" and stable hydrodynamic extensions, and makes the computation of the velocity field feasible.

Using this flow chamber and the built-in fluidic control it was possible to analyze samples at flow rates extending over 3 orders of magnitude : from 0.625 up to 625 $mm^3.s^{-1}$, i.e. with a mean velocity between 0.01 and 10 $m.s^{-1}$ in the sensing region. The only additional modification consisted in the conditional insertion of a flow restrictor into the sheath line upstream of the flow chamber to have a better control of low flow rates (Amblard et al. 1992).

7.3.2 Detection Optics

There was no need to modify any of the parameters of the standard optical configuration of the FACSTAR Plus™. However, the ability of the flow cytometer to distinguish two adhering cells from two cells that have been

---

[11] Illumination and optical detection take place in the straight channel using the same optical set-up as that of the standard FACSTAR Plus™.

[12] By "hydrodynamically equivalent" we mean that the fluid velocity in the vicinity of the axis through the actual chamber is identical to that in the virtual mode funnel. The equivalent radius b(z) is defined by $\pi b^2 = A(z)$, where A(z) is the area of the cross-section of the actual channel.

detached in the high stress region and closely follow one another through the sensing region strongly depends on the size of these cells relative to their mutual distance and to the dimension of the focused light source. In other words, the laser beam of the FACSTAR Plus™, can only resolve separated cells from attached cells if their separation is larger than the beam waist[13] in the flow direction. A 20 μm beam waist was satisfactory, but the space resolution could be improved by the use of a smaller waist.

7.3.3 Electronic Modifications

Since commercial instruments are built to be operated with high and quasi-fixed particle velocity in the sensing region, their detection electronics is adapted to process fast signals with typical duration in the 10 μs range. As expected, because reduced flow rates generate longer pulses in the 100-1,500 μs range, the detection circuitry was adapted to a wide range of temporal conditions, by modifying the relevant time constants as already explained (Amblard et al. 1992). Through a remote control box, the user was able to immediately adapt the time constants to the pulse length seen on the oscillographic display. Consequently, the regular analysis capabilities of the FACSTAR Plus™ were transferred over the whole range of flow rate.

Low velocity operation also raises other difficulties such as those related to increased duty cycle and baseline restoration in the detection circuitry. In our past work, these questions have been treated by a minimal solution as already discussed (Amblard et al. 1992). and should be considered with more attention for the design of post-prototypic solutions.

7.3.4 Operational Set-up and Control

The regular procedure required to start-up any work session on the FACSTAR Plus™ is valid for the modified flow cytometer also : the geometric alignment of the flow chamber with respect to the laser beam and the optical detectors, using calibrated fluorescent microbeads. In the modified instrument, this procedure can be most conveniently carried out first at fixed (medium or high) flow rate. One must then check that the alignment obtained is valid throughout the whole flow rate range. This

---

[13] The waist of gaussian beams is the cross-section at the edge of which the intensity is $1/e^2$ of its peak value.

ensures that the flow rate does not influence the quality of the analysis, and also indicates that the sample stream keeps focused in the center of the sheath stream at any flow rate.

In addition to the regular start-up procedure, one must also calibrate the rate of the sheath flow. This parameter is controlled in the FACSTAR Plus ™ by a manual sheath manometer and monitored by an electronic pressure sensor combined with an analog LED display. To get a more accurate control, we directly used the voltage delivered by this sensor. At the beginning of each session, a calibration chart was rapidly established to translate the sensor voltage into a flow rate.

Another critical point is the control of the sample flow rate. Indeed, reliable measurements are based on a steady-state hydrodynamic focusing, which requires the sample injection rate to be kept very stable and proportional to the sheath flow rate. In addition, bursts of the sample stream could cause the coincident detection of multiple particles and thus prevent the correct identification of particles. Practically, the rate at which particles are detected[14] was strictly kept below a limit[15] which increases with the flow rate. This point is one of the most critical parts of the technique. Noteworthy and unfortunately, low flow rate is not compatible with the production of a jet-in-the-air, and thus not compatible with cell sorting.

7.3.5 How to Do it ?

The instrumental developments described here can be directly reproduced on a FACSTAR Plus™ with a soldering iron, a few capacitors and resistors, wires, and a handful of switches and connectors ! Information concerning the manufacturing of the quartz flow chamber as well as the electronic modifications is available from the author. Nevertheless, since biology laboratories often do not have easy access to any engineering facility to build such modifications, the easiest solution would be that the developments presented here be made available from Becton-Dickinson as a kit. Other flow cytometers such as the Coulter ELITE™ can theoretically

---

[14] As displayed by the instrument with a one second refresh time.

[15] This upper limit was set to minimize coincidental or overlapping detection of independent particles. It was computed from a Poisson statistics, as a function of the duration of the pulses produced by the photodectors (Amblard et al. 1992).

be the basis of similar developments, but we have not practically investigated this possibility.

Because of the high price of flow cytometers such as the one we used, skillful investigators who have no need for a sophisticated multiparametric flow cytometer, could also consider the possibility to build a simplified flow cytometer dedicated to simple analyses at various flow rates and to adhesion measurement.

7.4 Interpretations and Possible Applications

7.4.1. Fluid Mechanical Interpretation of Dissociation Curves

The problem here is to elaborate a quantitative treatment of the dissociation curves such as the one shown on Fig. 2. Basic assumptions here are that the overall fluid flow is axisymmetric, steady-state and laminar[16]. Thus, the stress experienced by particles flowing along the axis is an axisymmetric extension the rate of which equals the gradient of the fluid velocity along the axis. One must first computes the overall velocity field in the whole chamber, then its value $v(q,z)$ on the axis, and finally $G(q,z)$ as the first z-derivative of $v(q,z)$. When applied to the critical flow rate $q=q_c$ extracted from the dissociation curves (Fig. 2), this procedure gives the distribution of the extension rates $G(q_c,z)$ which cause disruption of 50% of the particle doublets. If a doublet is broken by the series of extensions $G(q_c,z)$, it implies that it did not resist its maximum value. The $G(q_c,z)$ distribution can thus be represented by its maximum:

$$G_c = \text{Max}_{/z} \, G(q_c,z) = G_m \, (q_c) \qquad (4)$$

The above treatment is necessary to derive absolute gradient values, and the main practical problem it raises is the computation of the fluid velocity in the whole chamber as a function of the precise geomety of the chamber, which must be performed for each new chamber. In the case of our geometry, we were able to show that the Poiseuille approximation yields a very good approximation of $G_m(q)$ even if it is not valid for computing the velocity field (Amblard, 1993b). Therefore, the practical task consisted in

---

[16] This assumption is correct with respect to the maximum value of the Reynolds number evaluated in the flow chamber when operated at the highest flow rate : Re<2500.

(1) measuring the chamber geometry by microscopic observation, (2) computing the equivalent cross-section radius $b(z)$, (3) computing the axial Poiseuille velocity $v(z,q)$ and its derivative $G(q,z)$ through equations 2 and 3, and finally $G_m(q)$. If these numerical computations look too painful, one can completely skip them, and use disssociation curves and q_c values as a way to assess adhesion strength in a linear relative fashion which remains accurate nevertheless. For the flow chamber we used, the maximum of the velocity gradient ranges from 44 to 44,000 s^{-1} when q is between 0.625 and 625 $mm^3.s^{-1}$.

The mechanical effects of the pure axisymmetric extension experienced by paired particles have been extensively described in a companion paper (Amblard 1993b ; Fig. 3) :

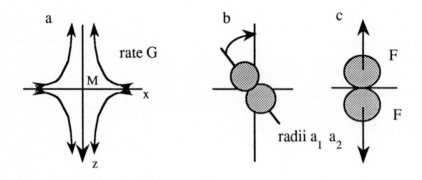

Fig. 3. Effects of the axisymmetric extension. Axial cross-section of the axisymmetric stream lines of the fluid velocity relative to the particle-fixed reference frame moving with a point **M** (**a**) along the central z-axis of the flow chamber. The orientation and extension effects are due to the viscous forces generated by the relative fluid motion (**b, c**).

- it stretches the fluid in the direction of the flow and compresses it in the perpendicular plane,
- it efficiently orients spherical objects "parallel" to the flow axis (Fig.3b),
- it tends to separate attached object oriented along z by an axial disrupting force (Fig. 3).

We found that the orientation efficiency only depends on the geometry of the chamber, and that pairs of rigid spheres with comparable sizes get fully oriented when entering the high stress region[17] (Amblard 1993b). In addition, we computed the vorticity diffusion length in water during the duration of the maximal stress[18]. At low and medium flow rates (in the two lower decades of our practical range) this diffusion length was at least one order of magnitude larger than the size if the biological cells we used (lymphocytes). Therefore, the continuous set of extensions was experienced in a quasi-static fashion by cells, and could be represented as a single brief[19] steady-state axisymmetric extension with a rate $G_m(q)$.

If the sample particles are rigid touching spheres, the critical extension G_c can be translated into a bulk disrupting force as already described (Amblard 1993b, Amblard et al. 1994) and given by:

$$F = 4\pi \mu \, a_1 a_2 \, G_c \, L \qquad (5)$$

where a_1 and a_2 are the radii of the particles, μ the fluid viscosity, and L a dimensionless function of the ratio a_2/a_1 such that $L(1)=1$. The hypothesis on the rigidity of the particles is crucial both to compute the orientation efficiency and to translate G_c into a bulk extension force[20].

Importantly, the comparison of the adhesion strength assessed here with forces measured by existing techniques is not meaningful because the force required to break adhesive contact is very dependent on the geometry and on the duration of the stress (Bell 1978), and the present technique is based on a unique axisymmetric and very brief extension which no other technique uses. As such, this approach is an isolated one whose results cannot be independently checked as yet. In addition, the information provided by the present method is given in a statistical manner : mean critical extension rate, mean adhesion strength, and nothing can be known about the geometry of the cell-cell contacts and how they are broken. These are important drawbacks of the approach described here.

---

[17] Defined as the region where $G(q,z)$ is larger than $G_m(q)/2$.

[18] Time spent in the high stress region.

[19] This time was in the 10 millisecond range in our experiments.

[20] The deformation of a particle sitting in an extension with rate G_c can be estimated by comparing its visco-elastic characteristics with the viscous pressure $2\mu G_c$ and the duration of the stress.

7.4.2 Assessment of the Adhesion Strength Between Leukocytes

The present work originates from an effort aimed at understanding which adhesion molecules were effectively involved in the adhesion between T and B lymphocytes during their interactions, and how much each contributed to the adhesion strength (Amblard et al. 1994). Cell-cell conjugates were made by mixing suspensions of intracellularly stained T and B lymphocytes, eventually in the presence of saturating concentrations of monoclonal antibodies which inhibit the formation of conjugates. Conjugate suspensions were then analyzed in the modified FACSTAR Plus™ at various flow rates, and dissociation curves were established. Using the theoretical tools described above, we were able to compute the critical extension rate in various circumstances. Adhesion strength between T and B lymphocytes was assessed and the strength contributed by the two major adhesion pathways involved in the interaction between human resting T and B lymphocytes was resolved. We found forces values between 0.15 and $0.3 \ 10^{-9}$ N (Amblard et al. 1994). The range of disruption forces that could be imparted upon T and B lymphocytes throughout the whole range of flow rate extended roughly from 3.10^{-11} to 3.10^{-8} N ; this range comprises the lowest cell-cell adhesion forces measured so far (Bongrand 1988; Baltz 1990). Virtually, future instrumental designs with "smoother" geometries could be useful to perform measurements over a lower range of forces.

7.4.3 Advantages and Potential Applications

Mechanical information provided by the present method can usually be obtained by various methods, in which the disrupting force is caused either by micromanipulation, hydrodynamic shear, or centrifugal force (Bongrand 1988). The unique advantages of using a flow cytometer are the following:

- the flow cytometer is the most powerful instrument for rapidly counting individual and associated cells[21],

---

[21]   Most other techniques (flow-induced detachment assays or micromanipulation) are very slow because they are based on the visual observation and the counting of individual conjugate-disruption events. In our previous work (Amblard et al. 1994), the lowest rate at which cells are analyzed was 50 s^{-1}.

- hydrodynamic focusing makes every individual cell experience the same extension,
- the disrupting stress is much shorter than in other techniques[22],
- the intensity of the extension can be much smaller than the minimal stress usually developed in micromanipulation techniques at equilibrium[23],
- their is no practical lower or upper limit for the size of particles, as long as the instrument can detect them.

Because the present approach is many orders of magnitude more rapid than any other previous techniques, it was possible for the first time to assess cell-cell adhesion forces in different biological circumstances, and in particular to use monoclonal antibodies against cell surface adhesion molecules to dissect quantitatively the molecular basis of these forces (Amblard et al. 1992). Moreover, with the high counting rate of flow cytometers, one can investigate conjugate formation even when this process has a low efficiency. Indeed, even a few percent of conjugates can be detected and mechanically assessed ; this is impossible for instance with micromanipulation methods which are too slow.

At last, the present flow cytometric disruption assay can be virtually combined with any other flow cytometric measurement: distribution of surface antigens relative to the contact area, calcium flux or various biochemical activation signals. Such approaches could give new insights into the physiology of cell-cell adhesion, and interesting information for conjugates formed between heterogeneous cell populations. Multiple applications of the present technique can be envisioned for the study of the molecular and mechanical phenomena underlying cell-cell adhesion and contact mediated intercellular recognition.

---

[22] So that the visco-elastic deformation of most nucleated cells is very limited (Schmid-Schönbein 1990) and one can assume that most of the disrupting force is mechanically transmitted to the contact region, with minimal viscous dissipation.

[23] Cell micromanipulation involves traction forces generated by small aspiration pressures of at least 1 mm H_2O. This is much higher that the lowest bulk extension pressure ($2\mu G$) that could be reached in our experimental conditions. But see Chap. 9.

References

Amblard F (1993a). Fluid mechanical properties of flow cytometers and assessment of cell-cell adhesion forces. In: Jacquemin-Sablon A and Crissman H (eds) Flow cytometry New Developments. ASI Series, Springer Verlag, Berlin p205

Amblard F (1993b) Hydrodynamic extensions applied to the assessment of contact forces between suspended particles: a theoretical description. submitted.

Amblard F, Cantin, C, Durand J, Fischer A, Sékaly R, Auffray C (1992). New chamber for flow cytometric analysis over an extended range of stream velocity and its application to cell adhesion measurements. Cytometry 13:15-22

Amblard F, Auffray C, Sékaly R, Fischer A (1994). Molecular analysis of antigen-independent adhesion forces between T and B lymphocytes. Proc Nat Acad Sci USA 91:3628-3632

Baltz J M, Cone R A (1990). The strength of non-covalent biological bonds and adhesion by multiple independent bonds. J Theor Biol 142:163-178

Batchelor GK (1967). An introduction to fluid dynamics. Cambridge University Press, Cambridge

Beug H, Katz FE, Stein A and Gerisch G (1973). Quantification of membrane sites in aggregating "Dictyostelium discoideum" by use of tritiated univalent antibodies. Proc Nat Acad Sci USA 70:3150-3154

Bongrand P (ed) (1988). Physical basis of cell-cell adhesion, CRC Press, Boca Raton

Crosland-Taylor PJ (1953). A device for counting small particles suspended in a fluid through a tube. Nature 171:37-38

Gerisch G (1980). Univalent antibody fragments as tools for the analysis of cell interactions in Dictyostelium. Curr Topics Dev Biol 14:243-249

Happel J, Brenner H (1983). Low Reynolds number hydrodynamics, Kluwer Academic publishers, Dordrecht Boston London, p 96

Kachel V, Fellner-Feldegg H, Menke E (1990). Hydrodynamic properties of flow cytometry instruments. In: Melamed MR, Lindmo T, Mendelsohn ML (eds) Flow cytometry and sorting. Wiley and Sons, New York, p 27

Melamed MR, Lindmo T, Mendelsohn ML (eds) Flow cytometry and sorting (1990). Wiley and Sons, New York

Pinkel D, Stovel R (1985). Flow chambers and sample handling. In: Van Dilla M A, Dean P N, Laerum O D, Melamed M R (eds) Flow cytometry: instrumentation and data analysis. Academic press, p 77

Schmid-Schönbein G (1990). Mechanical properties of leukocytes. Cell Biophysics 17:107-135

Thiery J-P, Brackenbury R, Rutishauser U and Edelman GM (1977). Adhesion among neural cells of the chick embryo. II. Purification and characterization of cell adhesion molecule from neural retina. J Biol Chem 252:6841-6846

Van Dilla M A, Dean P N, Laerum O D, Melamed M R (eds) Flow cytometry : instrumentation and data analysis (1985). Academic press.

White FM (1991). Viscous Flows. McGraw-Hill, New York

8 A Quantitative Cell Adhesion Assay Based on Differential Centrifugation of Bound Cells

P. Naquet

8.1 Introduction

Monitoring cell-cell interactions remains one of the critical steps in exploration of common biological systems. Control of cell adhesion is involved in embryogenesis allowing cells to segregate from each other to form distinct organs. Controlled adhesion is also involved in tissue repair and, as discussed below, in immune responses. Indeed, immune cells participate in the maintenance of cell integrity in the whole organism and thus permanently recirculate. Their adhesiveness is under the tight control of antigen-driven stimulus as well as signals coming from the microenvironment of the cell. Changes in adhesion properties are a consequence of this sensory function of immune cells.

Experimental approaches of cell adhesion are confined between two extreme situations : on one side, the interaction between complex cell membranes involving multimolecular complexes can be grossly evaluated as a global change of cell behaviour ; alternatively, purified or recombinant adhesion molecules are not always available to precisely quantify their involvement in a given interaction. Current methods for measuring cell adhesion are often poorly quantitative and do not allow the detection of subtle variations in adhesion properties of a given cell type in vitro. More sophisticated approaches involving flow cytometry or flow chamber have undoubtedly shed considerable light on mechanisms by which cells interact but require specific equipments. The purpose of this presentation is to present a simpler method which could be used as a starting approach in studying cell-cell interactions. Indications, limitations and possible improvements of the technique will be discussed through simple examples. The assay was first conceived by McClay to study the interaction between embryonic retina cells and monolayers of cells issued from distinct parts of the retina. This study provided a precise evaluation of adhesion parameters such as kinetics and strength of cell adhesion.

8.2 The McClay Assay

8.2.1 Principle

This technique is designed to study the interaction between a cell suspension and an immobilized cell or protein layer (prepared in a 96-well plate) and offers two main advantages :

- It is miniaturized and sensitive enough to handle simultaneously several experimental conditions.
- It avoids uncontrollable washing steps to separate bound from unbound cells since both cell populations are separated by upside down centrifugation of the plate.

Its main drawback is due to technical tricks leading to small variations of the results. The McClay assay might not be as useful for the exploration of homotypic interactions between non adherent cells.

8.2.2 Experimental Protocol

As described in the original reference (McClay et al. 1981), cell monolayers are cultured on the bottom of 96-well plates. Chromium-labeled retina cells (^{51}Cr) are added on top of the monolayer at 4°C in order to minimize variations before starting the assay. Then, the wells are filled up to the rim with cold medium to give a convex meniscus, and sealed with adhesive tape ; the plates can be incubated at different temperatures for different lengths of time (if a long incubation time at 37°C is required, it will be best to seal the wells at the end of the incubation). Conventional adhesion assays use washing steps with medium that considerably limit their reproducibiliy. Even when cautiously done, the introduction of a mobile phase creates turbulences in the well that are strong enough to break weak intercellular contacts. Thus, only strong interactions can be studied. In addition, it is impossible to rely on reproducible detachment forces since the washing strength varies from well to well. The McClay assay offers a solution to this problem. After sealing, plates are inverted at the end of the incubation time and spun down at chosen speeds in a centrifuge ; the force applied to bound cells is related to the centrifugal force applied to the plate (minimal dissociation can be applied by inverting the plate and letting only gravity (1 g) to detach the cells). Recovery of cell counts corresponding to the bound population involves a subsequent step of quick freezing of the plate. The use of plates made with flexible plastic allows one to cut the bottom of each well with a razor blade or a cutter when still frozen and to transfer this part of the well containing the adherent cell fraction to a separate tube for counting.

8.2.3 Experimental Models

Using this assay, McClay and colleagues analysed the parameters of retina cells adhesion to various cell monolayers issued from distinct regions of the retina (McClay et al. 1981). The biological question asked in this work was the understanding of topographic specificities of retina cells for distinct complementary cell populations present in the retina, controlling the precise spatial arrangement of neurons in the eye. Two main results were obtained : first topographic specificities in cell-cell interactions were identified ; second, temporal regulation of cell adhesion was dissected into an initial contact occuring at 4°C and a time and temperature-dependent reinforcement step. In each case, adhesion strengths were measured documenting a 2- to 13-fold amplification during reinforcement of cell adhesion under distinct experimental conditions.

Similar studies in other biological systems focused on the interaction of either embryonic cells, fibroblasts or glioma cells with components of the extracellular matrix (ECM), echinonectin and fibronectin or tenascin respectively (Lotz et al. 1989 ; Alliegro et al. 1990). In these studies, adhesive sites were studied using biological competitors or monoclonal antibodies. Again, initial or late adhesion parameters (kinetics, strength) were compared and correlated with the engagement of cell metabolism or cytoskeletal reorganization. Cells first bound to ECM molecules with a weak avidity and the interaction was reinforced within 15 minutes. These studies nicely correlated with others done on receptors of extracellular matrix molecules known as integrins. As recently reviewed by Hynes (1992), integrin molecules modulate their adhesive property as a function of time and cell activation. Integrins represent the predominant adhesion molecules involved in the stabilization of adhesion prior to migration and in the connection of extracellular signals to cytoskeleton.

8.3 Cell Adhesion During T Lymphocyte Maturation

8.3.1 Rationale

The initial assay was designed to probe interaction between homogeneous cell populations and a coated layer of matrix proteins or cells. The assay was adapted to the investigation of adhesion between a monolayer of adherent cells (in our case epithelial cells) and heterogeneous cell suspensions (thymocytes) (Lepesant et al. 1990). Controlling cell adhesion in time and space provides an easy way to sort cells from a complex population and thus to facilitate the temporospatial organization of the body. Our experimental model was applied to the intrathymic process of T lymphocyte maturation. In this system, cell differentiation is accompanied by migration of thymocytes from one stromal compartment to another, each providing stage-specific signals for maturation. It was therefore of interest to identify some of the molecules involved in the sorting of cells at different stages of differentiation.

8.3.2 Experimental Protocol

Material
Flexible 96-well plates and adhesive tape are obtained from Dynatech. A refrigerated table top centrifuge accepting plates should allow a precise setting of the centrifugal speed in order to provide reliable assays. Evaluation of precise g forces requires appropriate measurements in each centrifuge.

Preparing the plates
The Dynatech 96-well plates can be sterilized with ethanol and coated with various molecules to favor cell growth. We routinely used type I collagen (30 µg/ml) for at least 3 hours up to overnight incubation before growing cell monolayers. Since the quality and the reproducibility of the results rely on the quality of the monolayer, a special care should be taken in terms of cell numbers and homogeneity of the monolayer (rather use cells with contact inhibition than tumors forming several cell layers). A strongly adherent cell is optimal for this type of assay and we would not suggest to use this assay to probe adhesion between poorly adherent cells. A precise evaluation of the stability of the monolayer when applying centrifugal forces should be done before adding other cell types. In our hands, cell monolayers remained strongly attached to the bottom of the

wells beyond 800 g (upper limit of our centrifuge).

Cell labeling
Cell labeling with ^{51}Cr requires at least 3 hours of incubation of cells in a small volume containing roughly 300 µCi of ^{51}Cr.

Assay
Plates, cell suspensions and medium are kept on ice before starting the assay. The volume of medium in each well should be at least 150 µl before adding cell suspensions in a smaller volume. Then, several possibilities are offered. Cells can be shortly and smoothly spun on top of the monolayers (10 to 30 g for 3' for example) or simply allowed to settle at 1 g. In every case, the temperature should be kept as close as possible to 4°C to allow a precise timing of the experiment. Once the cells have reached the bottom, the top of the plate is dried and the wells (almost full at this stage) are filled up to the rim (adding slowly cold medium does not seem to affect the binding of the cells in the bottom of the wells) to give a convex meniscus. Special care should be taken to avoid wetting the top of the plate. Then, the adhesive tape is rolled on top of the plate, carefully avoiding the formation of air bubbles. Wells must be sealed without major contamination from one well to the other when one rolls the adherent tape on top of the plate. Thus, the assay should be designed leaving one free row of wells between each row of the assay and at the end of the plate. Plates can be incubated for different lengths of time at different temperatures before sealing the plates, and centrifuging upside down for at least 5 minutes. Some of us have also used a moderate rotation to keep cells in movement for the initial contact; this specific experimental condition might be useful to detect adhesion processes involving selectin molecules known to participate in the rolling of circulating leucocytes on endothelial cells. Provided that these conditions are respected, the assay will be reliable ; these technical aspects require some practise. We routinely tested quadruplicate samples (sometimes more) to ensure the reliability of the assay. The end of the assay precisely follows the initial experimental conditions described by McClay.

Binding strength calculation
The formula for evaluating the dislodgement force is the one used by McClay :

$$Fd = (\rho_{cell} - \rho_{medium}) \times V_{cell} \times RCF \times 1\ g \qquad (1)$$

in which Fd is the dislodgement force per cell in dynes, ρ_{cell} is specific density of the cell (1.07 g/cm^3), ρ_{medium} is specific density for medium (1 g/cm^3), V_{cell} is cell volume in cm^3 (to be adapted to the diameter of each cell type), RCF the radial centrifugal force obtained in the centrifuge (given in g = 981 cm/s^2) applied to the inverted plate. Results are expressed in dynes per cell.

8.3.3 Experimental Model

The biological question
In the thymus, epithelial cells form a complex network through which thymocytes are selected and differentiate. Two major compartments (cortex and medulla) containing distinct microenvironments condition full T cell differentiation and provide stage-specific signals delivered via soluble or membrane molecules. Differential adhesion properties allow thymocytes to move from one microenvironment to the other ; it was of interest to explore which molecules controlled the interaction between each cell type at different stages of maturation.

Results
The interaction between maturing thymocytes and thymic epithelial cell monolayers was analyzed. Thymocytes are heterogeneous and as shown in Fig. 1a, a progressive dissociation of bound cells was observed when increasing the dissociation force. However, when homogeneous cell clones are used, progressive dissociation may also be seen, suggesting that individual variations from cell to cell or technical inaccuracies (due to lack of synchronization of cells or heterogeneity of the cell monolayer) are involved. This result is nevertheless highly reproducible and allows the discrimination between strongly and poorly adherent cell populations. By raising the temperature up to 37°C and incubating cell cultures for 30 to 60 minutes, an increase in the percentage of bound cells and a reinforcement of cell contacts requiring stronger dissociating conditions were noticed (Fig. 1b). It is worth mentioning that even at the highest centrifugal speeds, there were still some undetached cells and cell monolayers were unaffected. The effect of various monoclonal antibodies specific for thymocyte-specific membrane molecules were then tested for their inhibitory effect on cell adhesion as a function of time (Fig. 1d and f) or centrifugation speed (Fig. 1c and e). To indirectly evaluate the involvement of a given molecule in the stability of the interaction, we reasoned that if the inhibitory effect of a given mAb was still detectable at high dissociating speeds, it meant that the interaction mediated by the

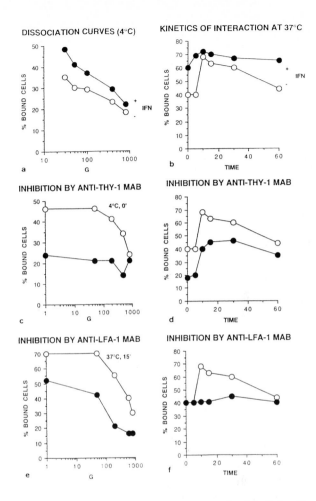

Fig. 1a-f. Quantitative and qualitative parameters in thymocyte / epithelial cell interaction .Results are expressed as percentages of bound cells as a function of time or centrifugation speed. Results in **a** and **b** show dissociation curves of adult thymocytes as a function of G or time with epithelial cell monolayers treated (*dark symbols*) or untreated (*open symbols*) with IFN for 3 days prior to the assay. Inhibition experiments with (*dark symbols*) or without (*open symbols*) mAb correspond to distinct experimental settings in order to maximize the effects of the mAb and are also shown as a function of time (**d** and **f**) or G (**c** and **e**). Antibodies were preincubated with thymocytes and left throughout the adhesion assay. Anti-Thy-1 mAb mainly block the initial contact at 4°C whereas the effect of anti-LFA-1 mAb is optimal after 15' at 37°C. IFN was added to monolayer cultures 48 hours before the assay. These results are extracted from distinct experiments and in some cases have been simplified.

molecule recognized by this mAb was of strong avidity. Although both Thy-1 and LFA-1-specific mAb partially inhibit cell adhesion at low centrifugation speed, only LFA-1-specific mAb interfere with residual cell interaction at high speed. All these results can be summarized as follows : a significant fraction (30%) of murine thymocytes bound to uninduced thymic epithelial cells. The percentage of bound cells varies with the age of the animal and is significantly enhanced when epithelial cells are pretreated with interferon (IFN) known to induce high levels of expression of several adhesion molecules on stromal cells. Secondly, the bound thymocytes predominantly belong to a subset of thymocytes identified by the expression of the CD4 and CD8 molecules. This subset is in tight interaction with epithelial cells in vivo and thus this result nicely reflects the physiological situation. Thirdly, among the bound thymocytes, one can differentiate two adhesion mechanisms. The majority of bound cells interact with the epithelial cell monolayer with a low avidity since these cells can be dissociated at low centrifugal forces corresponding roughly to adhesion strengths of the order of $10^{-3/-4}$ dynes per cell (equivalent to 1000-10000 piconewton). This calcium and temperature-independent interaction was found to involve the Thy-1 molecule (Fig. 1c and d) (Hé et al. 1991). In a more recent study, the results obtained in the Mc Clay assay were confirmed using a flow chamber (Hé et al., unpublished results).

Another fraction (20%) of thymocytes showed a distinct behavior after interaction with the epithelial monolayer. As also observed in the study by McClay, the initial contact of these cells was followed by a reinforcement of cell adhesion with adhesion strengths increasing up to 10^{-2} dyne per cell. The reinforcement was time- (15-30') and temperature- (37°C) dependent and also involved cell metabolism (Fig. 1b). Finally, this reinforcement was found to be transient, lasting for about one hour. These cells belonged to a minor subset of CD4+CD8+ thymocytes expressing low levels of CD3-T cell receptor molecules. The reinforcement but not the initial Thy-1-dependent step could be inhibited by LFA-1-specific mAb, another member of the integrin family interacting with a cell-bound ligand known as ICAM-1 expressed by the thymic epithelial cell line used (Fig. 1e and f). Our observation fitted nicely with another work on the function of the LFA-1 molecule (Dustin and Springer 1989). In this report, the authors had shown that the affinity of LFA-1 for purified ICAM-1 molecules increased rapidly (15') after cell activation but lasted for only a short period. Thus, these results on complex cell populations correlated with those obtained in a reductionist system, supporting the notion that this assay can be used to rapidly probe complex kinetics in cell-cell interactions without initially relying on purified molecules or more sophisticated experimental designs.

8.3.4 Trouble-Shooting

This assay seems ideal at first sight. However, we have reproducibly encountered distinct problems essentially due to the technical design. These problems will be described as they appear in the experimental set-up.

The type of plates used for the assay is not optimal : indeed, since the end of the assay requires cutting of the bottom of the wells, one needs to use flexible plates. These plates present two major artifacts ; first, they are not treated for cell culture and even after treatment of the wells with different molecules to increase monolayer cell formation (poy-L-lysine, collagen, etc.), the quality of the coating is never optimal, leading to irregular monolayers. This problem creates significant variations among identical replicates and is only solved by increasing the number of experiments and replicates, thus severely limiting the use of the technique in a large scale approach which was one of the initial reasons for using it. Second, the bottom of these plates is not flat and thus the force applied to each cell on the bottom is not strictly identical on the edge and in the center of the well. Although this variation is of little amplitude, it may also limit the accuracy of measurement of adhesion strengths.

The setting of the assay needs to leave free rows between experimental points to avoid contamination from one well to the other; however, if cells are already at the bottom of the plate before sealing, the overflow of medium does not lead to important variations of the results.

The formation of air bubbles is a constant problem but can be well controlled if one uses solid and not flexible plates ; then recovery of bound cells has to be adapted. McClay and colleagues proposed an alternative technique to overcome this problem by simply spinning a non-sealed plate on top of a symmetrical plate filled with medium. We have tested this technique and found it tricky. However, it allows recovery of both bound and unbound cells.

The cell-labeling technique may also be a source of artifacts. Indeed, a precise quantification of bound cells should be totally independent of cell-specific parameters modifying the quality of labeling. The use of ^{51}Cr still represents a reasonable choice when one deals with homogeneous cell populations but may be a source of inaccuracy if the cell population studied is heterogenous as in our experimental system. The ideal situation would be to use a nuclear staining as the most accurate estimation of cell numbers. Such techniques exist and are mainly based on agents intercalating with DNA and detectable by fluorescence (Hoechst, for example). The use of this technique would considerably simplify the assay provided that the read-out technique is accurate enough.

8.3.5 A Modified McClay Assay

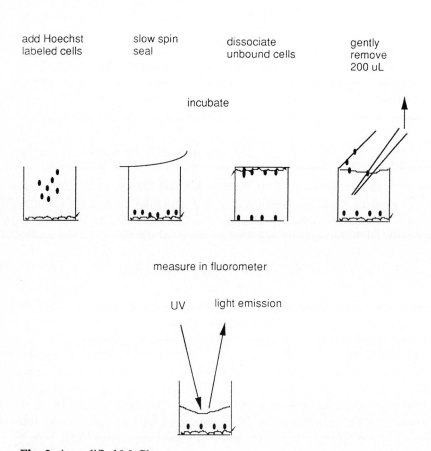

Fig. 2. A modified McClay assay

For all these reasons, we would suggest to improve the current assay as indicated in Fig. 2 :

- cell staining with Hoechst (a viable DNA stain);
- setting the cell monolayers or matrix proteins in conventional culture-treated 96-well plates, leaving a free row every other row ;

- proceeding as usual for the initiation of the assay and sealing of the plates (plates with individualized wells could be used) ;
- after the upside down spinning to separate unbound cells, individual plates should be kept on ice ; one after the other, they should be briefly put back in the normal position, the adhesive tape quickly pealed and using a multichannel pipette, the 200 upper microliters of individual rows removed ; this part can be safely done within a very short period of time to avoid sedimentation of few unbound cells back on the monolayers ;
- the remaining fluorescent cells could then be detected in a fluorescence detector using a UV light and adapted to the lecture of microtiter plates. This method is faster than counting individual wells in a gamma counter and in our preliminary experiments has proven to be as sensitive as, if not more than, the radioactive counting. However, our experience with this possible improvement is still in a preliminary stage. However, such equipments, although still expensive, are available and some also allow fluorescent detection at different wavelengths for excitation and detection, thus extending the type of fluorochrome used in the assay.

8.4 Conclusion

The McClay assay is a reasonable experimental choice if one is interested in the following questions :

- adhesion between adherent cell monolayers (or coated molecules) and cell suspensions (see as an example the summary in Fig. 3) ;
- quantification of adhesion strength and kinetics during the initial stages of interaction ;
- miniaturization of cell adhesion assays.

The main problems are essentially due to the technical design that introduces variability at different steps. However, the proposed modifications should largely simplify the procedure.

Acknowledgment. The author wish to thank Hélène Lepesant, Hai-Tao Hé and Anne Odile Hueber who participated to the progressive improvement of this technique and Pierre Bongrand for initiating the course and commenting this manuscript.

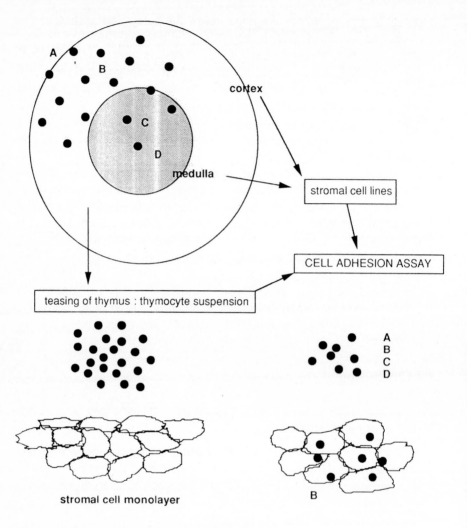

Fig. 3. Summary of the experimental set-up

References

Alliegro MC, Burdsal CA, McClay DR (1990) In vitro biological activities of echinonectin. Biochemistry 29:2135

Dustin ML, Springer TA (1989) T cell receptor cross linking transiently stimulates adhesiveness through LFA-1. Nature 341: 619-624

He HT, Naquet P, Caillol D, Pierres M (1991) Thy-1 supports adhesion of mouse thymocytes to thymic epithelial cells through a Ca^{2+}-independent mechanism. J Exp Med.173:515-518

Hynes OR (1992) Integrins : versatility, modulation and signaling in cell adhesion. Cell 69:11-25

Lepesant H, Reggio H, Pierres M, Naquet P (1990) Mouse thymic epithelial cell lines interact with and select a CD3lowCD4+CD8+ thymocyte subset through an LFA-1-dependent adhesion-de-adhesion mechanism. Int. Immunol. 2:1021-1032

Lotz MM, Bursdal CA, Erickson HP, McClay DR (1989) Cell adhesion to fibronectin and tenascin : quantitative measurements of initial binding and subsequent strengthening response. J Cell Biol 109:1795-1805

McClay DR, Wessel GM, Marchase RB (1981) Intercellular recognition : quantitation of initial binding events. Proc Natl Acad Sci USA 78:4975-4979

9. Picoforce Method to Probe Submicroscopic Actions in Biomembrane Adhesion

E. Evans, R. Merkel, K. Ritchie, S. Tha, A. Zilker

9.1 Introduction : Adhesion in Biology

Adhesive interactions are central processes in numerous biological functions like tissue assembly (development) and identification and removal of alien organisms in immune defense (Springer 1990 ; Takeichi 1991). It is important to recognize that the consequences of cell adhesion are more than surface bonding. For biological organisms, adhesion usually initiates signalling pathways to activate and modulate internal cell functions. These signals often lead to physical transformation of the cell interior so that it becomes more/less rigid, motive, or strongly linked to the substrate structure. Indeed, adhesion may result in complete integration of the cell interior into a macromaterial (i.e. tissue) where molecular stress-bearing linkages penetrate the cell membrane. Although the phenomenological features of cell adhesion are well recognized, little is known about the submicroscopic physical mechanisms that ultimately implement these important biological functions. Because the sites for surface bonding are sparsely distributed compared to the range of bonding forces, the actions are usually lost in a juggernaut of cell surface movements when contacts form (Evans 1994). Hence, the most common physical assay of adhesion is the strength of attachment. When intersurface bonds are very strong, the strength of attachment images the physical coupling of surface receptors to the cell membrane and cytoskeletal structure. On the other hand, if receptors are strongly anchored to the cell structure, the attachment strength may expose the receptor-ligand interaction. Thus, key questions to be addressed in physical tests of adhesion are : is adhesive strength governed by ligand-receptor bonds or linkages to cytoskeletal structure? Does receptor anchoring to the cytoskeletal structure change with binding different ligands to the receptor? What are the mechanisms involved when cells separate from substrates?

9.2 Macroscopic Adhesion Experiments

What is tested in cell adhesion experiments depends on professional orientation. For instance, biologists have usually relied on the phenomenology of stuck versus nonstuck to examine surface receptor expression, effects of ligand and locus of binding, and functional consequences in cell behaviour. By comparison, engineering and physical scientists have focused on macroscopic forces required to detach adherent cells ; the schematics in Fig. 1 demonstrate several types of physical detachment. As shown, the forces used to detach spread cells end up distributed by cell deformation over numerous molecular attachments - each experiencing a different level of force. In a few cases, the results have been digested to yield the membrane peeling tension that defines the mechanical energy used to "fracture" a unit area of adhesive contact (an example is shown in Fig. 2). Even with careful experimental design and analysis, macroscopic measurements of fracture energy involve many properties beyond that of bond dissociation (Evans 1994) : e.g. the density of linked receptors, the strengths of receptor-ligand bonds and receptor-membrane anchorage. Furthermore, smoothing of contact roughness and accumulation of linked receptors during contact separation are important dynamic processes that enrich the number of attachment sites at the contact. Even if thermodynamic properties of receptor-ligand binding are known, the density of attachments local to a contact and the locus of molecular failure remain indeterminant. Thus, new approaches are needed to expose the physical actions that govern adhesion at the submicroscopic level.

9.3 New Direction in the Study of Biological Adhesion

What scientists would like to know about cell adhesion again depends on perspective. In biology, the objectives are to describe the consequences of intersurface attachment on cell biochemistry and function (e.g. chemical

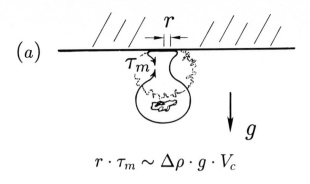

$$(a)$$

$$r \cdot \tau_m \sim \Delta\rho \cdot g \cdot V_c$$

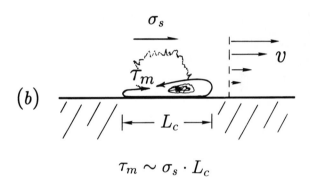

$$(b)$$

$$\tau_m \sim \sigma_s \cdot L_c$$

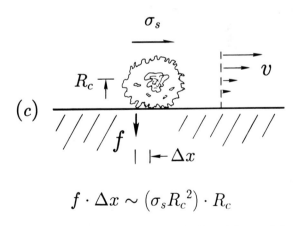

$$(c)$$

$$f \cdot \Delta x \sim \left(\sigma_s R_c^2\right) \cdot R_c$$

Fig. 1a-c. Detachment of soft-deformable cells **a** by gravitational force ; **b** by fluid flow. Peeling tension τ_m separates the cell from the surface. **c** Detachment of a stiff-spherical cell. Fluid shear acts on the cell through a rotational torque.

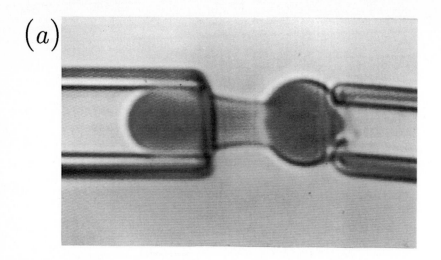

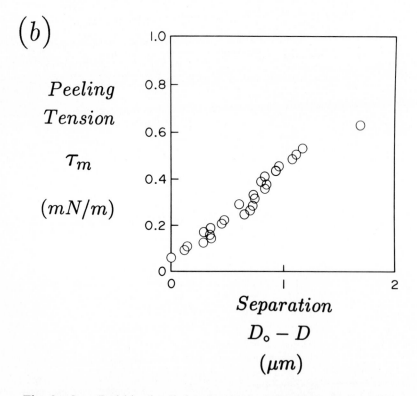

Fig. 2a, b. a Red blood cell detachment by micropipet suction (Evans et al., 1991 b). The deformable cell (*left*) was bonded to a chemically crosslinked cell (*right*) using monoclonal antibodies to surface glycoproteins. **b** The symmetrical shape and known properties of the red cell membrane lead to precise analysis of the peeling tension as a function of separation.

signalling and cytostructural modifications). In physics, the goal is to determine the range and dynamics of bond formation, strength of molecular attachment and locus of failure in relation to molecular physics and chemistry. Clearly, the objectives of both communities are valuable but new methods are needed to unify these objectives. To achieve this aim, we must recognize the criteria for design of a physical probe to test single bonds between soft-biological interfaces in liquid environments. These criteria are forces with magnitudes well below the strengths of ionic and covalent bonds as illustrated in the following logarithmic hierarchy :

$10^3 - 10^4$ pN   ionic - covalent bonds

$10 - 10^2$ pN   van der Waals - hydrogen bonds

1 pN   force necessary to stretch a 10 nm flexible polymer to twice its characteristic dimension. Discrete force produced by a myosin/actin molecular "motor"

< 0.1 pN   force necessary to displace a flexible membrane (tension ~ 10^{-5} N/m) outward by 10 nm

Since a piconewton (pN) force is equivalent to a 10^{-10} g weight, we need an extremely sensitive force "balance" to measure molecular forces at soft-biological interfaces!

Amongst surface probes, the benchmark for small force measurement has been the atomic force microscope AFM (Binnig et al. 1986 ; Weisenhorne et al. 1989 ; Maivald et al. 1991). The concept behind the AFM is a simple cantilever beam as sketched in Fig. 3. The mechanical stiffness k_f establishes the force sensitivity (force/tip deflection) as given approximately by :

$$k_f \approx E(t/L)^3 w/4 \qquad (1)$$

The elastic modulus "E" of the AFM cantilever material is ~ 5 x 10^{10} N/m^2; "w" is the width of the cantilever, "t" is the thickness, and "L" is the length. Very sensitive cantilevers have stiffnesses of order 10 mN/m which represents a 0.5 μm thick beam 150 μm long and 20 μm wide. Obviously, it is a major technological feat to fabricate long cantilevers with thicknesses of a wavelength of light ! With such a thin cantilever, a one "atom" deflection (0.1 nm) is equivalent to 1 pN force. However, this level of force is less than the resolution limit for the cantilever implied by thermal fluctuations :

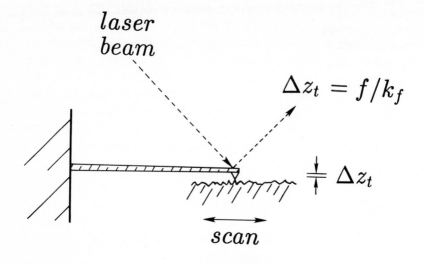

Fig. 3.   Schematic of the atomic force microscope (Binnig et al. 1986).

$$\Delta f ~\sim~ \sqrt{k_B T k_f} ~\sim 10 \text{ pN} \tag{2}$$

where $k_B T \sim 4 \times 10^{-21}$ N-m. Similarly, thermal fluctuations in position of the cantilever tip are scaled by $\Delta x ~\sim~ \sqrt{k_B T / k_f} ~\sim 1$ nm. Force detection can be improved beyond thermal resolution limits with the use of exotic techniques (Olsson et al. 1992) ; but these methods are difficult to implement for in liquid environments. Indeed, fluctuations of molecular structures at biological interfaces are likely to create large excursions of at least 1-10 nm, maybe more ! Clearly, an even softer transducer is desirable with a stiffness that can be tuned over a wide range of forces.

9.4 New Biosurface Force Probe

Motivated by AFM technology and guided by experience in micromechanics of biomembranes, we have developed an ultrasensitive-tunable force transducer that can measure forces over a range from 0.01 pN up to the strength of covalent bonds (>1000 pN). The transducer is simply a microbead probe attached to a pressurized membrane capsule (see scheme in Fig. 4). The pressure P is controlled

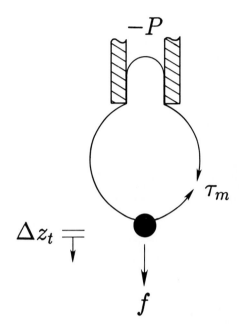

Fig. 4. Scheme of an ultrasensitive-tunable force transducer formed by a pressurized membrane capsule.

by micropipet suction and sets the capsule membrane tension τ_m :

$$\tau_m = P \cdot R_p/2(1 - R_p/R_0) \qquad (3)$$

R_0, R_p are the radii of the membrane capsule and suction pipet respectively. When a small force f is applied to the probe, the spherical shape of the capsule is elongated by a displacement Δz_t proportional to the force. The stiffness constant k_f for the transducer ($f = k_f \cdot \Delta z_t$) is given explicitly by :

$$k_f = 2\pi \cdot \tau_m/[\ln(2R_0/R_p) + \ln(2R_0/r_b)] \qquad (4)$$

r_b is the radius of the circular contact between the capsule and the microbead. Because the stiffness is proportional to tension, the force sensitivity can be tuned in operation between 1 μN/m and 10 mN/m simply by changing the pipet suction pressure ; no separate calibration is needed.

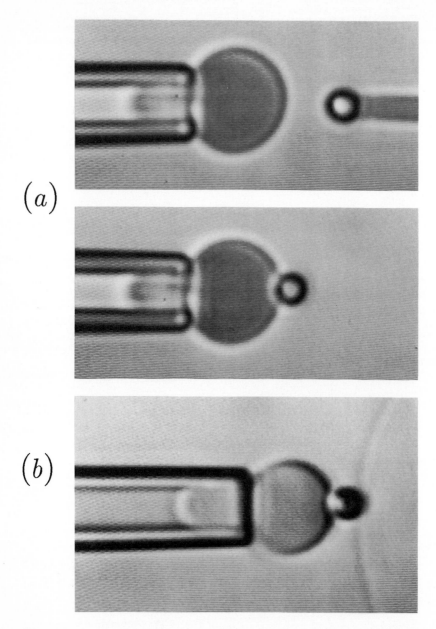

Fig. 5. a Microbead assembly to the transducer surface. **b** Probe attachment to a receptor in a biomembrane surface. The outer dimension of this red cell transducer is about 5 μm.

Displacements of the probe are readily measured down to values of 0.01 μm with optical techniques. Thus, the range of forces accessible to this simple probe is amazing, i.e. 1000 - 0.01 pN. For comparison, recall that rupture of a weak hydrogen bond requires 100 pN or more. With a similar transducer, we have shown that hydrophobic anchoring of receptors in the bilayer of a membrane is much weaker than a hydrogen bond, i.e.~ 10 pN (Evans et al. 1991 a).

A versatile procedure has been developed for coupling the transducer to arbitrary surface receptors. The solid microbead is chemically conjugated with separate ligands for macroscopic glueing to the transducer surface and focal bonding to receptors in a biological surface. Figure 5 shows assembly of the microbead to the membrane capsule and a prototype test of the probe. By reducing the amount of ligand specific to the biological receptor, a density level is reached where point contact between the probe and the surface results in infrequent-discrete attachments characteristic of single molecular complexes.

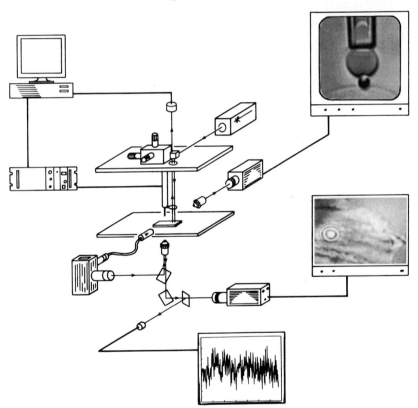

Fig. 6. The apparatus used to observe transducer placement and to monitor submicroscopic displacements of the microbead probe.

The apparatus for observation of transducer position and detection of nanoscale movements of the probe involves an orthogonal microscope system (Fig. 6). A standard inverted microscope is fitted with reflection interference contrast RIC optics to image the position of the microbead above the test surface and a horizontal microscope (oriented perpendicular

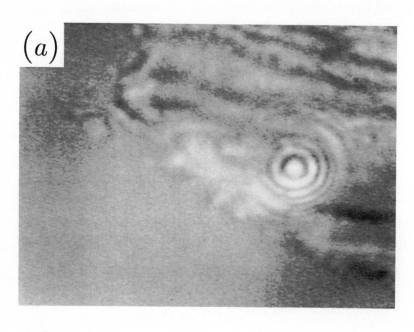

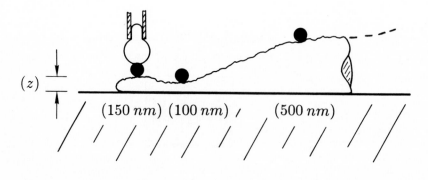

Fig. 7. a The RIC image of a microbead in contact with the upper surface of a fibroblast cell spread on the coverglass. **b** Cell thickness found at various

locations.
to the optical axis of the inverted scope) is used to macroscopically position the force probe. A piezoelectric device is actuated in series with course micromanipulation of the transducer for precise translation of the probe to and from the cell surface. Figure 7a shows an RIC fringe pattern produced by a microbead positioned on top of a biological cell ; Fig. 7b shows the apparent thickness measured under low force (~ 10 pN) at different positions on the cell. Microbead elevations above the coverglass are derived from the interference pattern following developments given in Zilker et al. (1992) and Radler and Sackmann (1993). A four quadrant photodetector is used to locate the center of the ring pattern where high resolution movements of the microbead are detected from intensity changes. To correct for thermal drift and hysteresis in the piezo translator, a laser interferometer is currently being assembled to monitor the relative position between the cover glass substrate and the piezo-micropipet junction. Distance adjustments will be fed back to control the transducer position. In situations where interactions between the microbead and test surface are uniformly distributed, this arrangement can be used as a microscopic version of the Surface Force Apparatus (Israelachvili and Adams 1978 ; Israelachvili 1985 ; Parker et al. 1989). In the future, the test platform will be translated by a two axis piezo to image the biological surface under low force with a vertical resolution of a few nm. Although many applications can be envisioned, the important question is how to take advantage of the ultrasensitivity of this transducer.

9.5 Exposing Submicroscopic Actions in Biological Adhesion

Although the tendency is to seek precise measurements of forces and positions, it must be recognized that the prominent features of submicroscopic actions at a biological interface will be strong fluctuations and marginal determinism ! Hence, information must be extracted from "noisy" data. Rather than cringe at this prospect, experiments should be configured to take advantage of fluctuations. Indeed, the study of fluctuations can provide significant insights into the range and statistics of single bond formation, the structural properties and modulation of receptor-cytoskeletal linkages, and the dynamics and failure of single attachments under stress. The following experimental protocols are being developed to expose these features as illustrated in Fig. 8 : to examine bond formation, the transducer stiffness is set low so that the probe exhibits major fluctuations in position ($\Delta z_b \geq$ a few nm) as it is moved

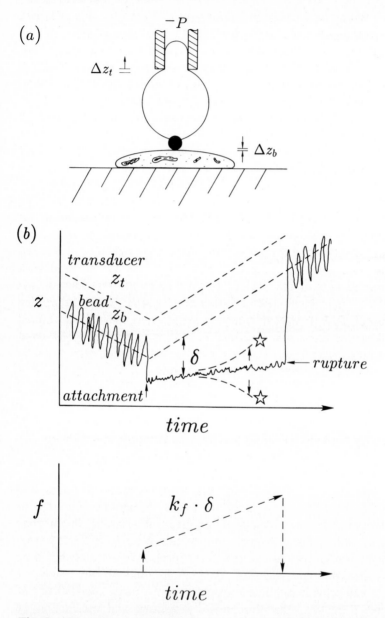

Fig. 8. a Transducer displacement (Δz_t) and microbead movement (Δz_b) to be monitored during an experimental test. **b** Submicroscopic response expected for the microbead as it approaches formation of an attachment - subsequent to attachment - and after rupture of the attachment. Deviations (*) reflect

receptor-substrate transformations signalled by bonding.

towards the biological interface. Bond formation is signalled when probe fluctuations diminish precipitously as a consequence of the added receptor-interface stiffness. Likewise, bond release shows up when probe fluctuations return to the original level set by the low transducer stiffness. The relative frequency of formation and release quantitates on/off rates at the submicroscopic level in the absence of competitive soluble ligands. Next, while bound, receptor-interface rigidity is tested directly by stressing the attachment (the effect of substrate rigidity on fluctuations of the probe is shown in Fig. 9), changes in interface stiffness subsequent to bonding will indicate structural transformations signalled by the bonding event.

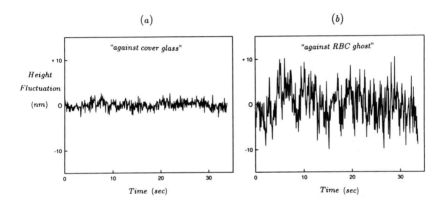

Fig. 9a, b. Fluctuations of probe position observed over time when the microbead is held against **a** the rigid coverglass substrate and **b** a soft red blood cell ghost under low force (< 10 pN).

Finally, the transducer is retracted to load the attachment until failure. As shown in the Appendix, study of attachment failure at different loading rates exposes intrinsic dynamics of the rupture mechanism. When attachments rupture, a major puzzle is often the locus of failure, e.g. ligand-receptor bond or receptor-cytoskeletal linkage. A useful assay of rupture site is to attempt reassembly of the attachment after rupture. If the ligand-receptor bond is broken, it is expected that the bond should reform upon repositioning the probe near the receptor but if the receptor is extracted from the membrane, there should be no reattachment (caution must be used here since unexpected lateral displacements of the probe could compromise the test). Even though subtleties and technical problems will be encountered, these protocols allow sensitive study of the stochastic features of bond formation and rupture on the submicroscopic scale.

9.6 Comments

The advent of atomic force microscopy has taught us that classical methods can be used to expose invisible textures and mechanical properties of molecular structures on the submicroscopic scale. AFM technology is rapidly developing with a major thrust in the direction of biological applications (Radmacher et al. 1992). By necessity, this thrust requires significant reductions in cantilever stiffness and close attention to tip-surface compatibility. Even though progress has been made, it is likely that new technology will be necessary for study of ultrasoft interfaces in biology. The approach described here is being developed for just this purpose. The expectation is that a whole new realm of physical study of biointerface processes will be enabled by introduction of ultrasensitive force methods.

References

Binnig G, Quate CF, Gerber CH. (1986) Atomic force microscope. Phys Rev Lett 56:930-933

Evans, E. 1994. Physical actions in biological adhesion. In : Lipowski R, Sackman E (eds) Biophysics handbook , Vol. I. Springer, Berlin (in press)

Evans E, Berk D, Leung. A (1991a) Detachment of agglutinin-bonded red blood cells. I. forces to rupture molecular-point attachments. Biophys J 59:838-848

Evans E, Berk D, Leung A, Mohandas N (1991b)   Detachment of agglutinin-bonded red blood cells : II. mechanical energies to separate large contact areas. Biophys J 59:849-860

Israelachvilli JN (1985) Intermolecular and surface forces. Academic Press, London pp 1-296

Israelachvilli JN, Adams GE (1978) Measurement of forces between two mica surfaces in aqueous electrolyte solutions in the range 0-100 nm. J Chem Soc Faraday Trans I 74:975-1001

Maivald P, Butt HJ, Gould SAC, Prater CB, Drake B, Gurley JA, Elings VP,

Hansma PK (1991) Using force modulation to image surface elasticities with the atomic force microscope. Nanotechnology 2:103

Olsson L, Tengvall P, Wigren R, Erlandsson R (1992) Interaction forces between a tungsten tip and methylated SiO2 surfaces studied with scanning force microscopy. Ultramicroscopy 42-44:73-79

Parker JL, Christenson HK, Ninham BW (1989) Device for measuring the force and separation between two surfaces down to molecular separations. Rev Sci Instrum 60:3135-3138

Radler J, Sackmann E (1993) Imaging optical thicknesses and separation distances of phospholipid vesicles at solid Surfaces. J Phys (Paris) II 3:727-

Radmacher M, Tillman RW, Fritz M, Gaub HE (1992) From molecules to cells: imaging soft samples with the atomic force microscope. Science 257:1900-1905

Springer TA (1990) Adhesion receptors of the immune system. Nature 346:425-434

Takeichi M (1991) Cadherin cell adhesion receptors as a morphogenetic regulator. Science 251:1451-1455

Weisenhorne AL, Hansma PK, Albrecht PR, Quate CF (1989) Forces in atomic force microscopy in air and water. Appl Phys Lett 54:2651-2653

Zilker A, Ziegler M, Sackmann E (1992) Spectral analysis of erythrocyte flickering in the 0.3-4 μm-1 regime by microinterferometry combined with fast image processing. Phys Rev A 46:7998-8001

Appendix

As developed previously (Evans 1994 ; Evans et al. 1991a), the statistics for failure of a molecular attachment under force f can be cast in terms of an underlying rupture strength f_0, an intrinsic frequency of failure υ_0 (#/time), and an exponent "a" that specifies the dynamic coupling of failure frequency to force, i.e.

$$\upsilon \approx \upsilon_0 (f/f_0)^a \qquad (5)$$

Simple analysis shows that the level of force "f" for most frequent rupture depends on rate of loading $\dot{f}$ (force/time) through,

$$"f" \sim f_0 (\dot{f} / \upsilon_0 f_0)^{1/a+1}$$

Hence, measurements of attachment failure at well controlled loading rates can lead to the intrinsic properties that govern the stochastic process.

10. Study of Cell-Cell Interactions with the Travelling Microtube

H. L. Goldsmith, K. Takamura, S. Tha and D. Tees

10.1 Introduction

We describe a microrheological technique developed to study the motions and interactions between colloidal particles or between biological cells suspended in fluids undergoing laminar viscous, Poiseuille flow through a circular tube. The basic principle of the method is to observe the translational and rotational movements of individual suspended particles, and their collisions with each other, through a microscope by tracking them in flow down precision-bore glass tubes of 50 - 250 µm diameter. This is achieved by moving the tube and the infusion syringe (or gravity feed reservoir), mounted on a hydraulically driven platform, in a direction opposite to that of the flow while focussing on a particular particle or cell. As an example, we show the two-body collision between equal-sized spheres as it appears when the particles are tracked by moving the tube upward with a velocity equal to that of the downward flowing fluid at the mid-point of the axis between the two spheres. The collision shown in Fig. 1 is one in which the trajectories of the spheres are symmetrical, as observed in the case of macroscopic uncharged particles in viscous media (Goldsmith and Mason 1964). The particles separate along paths having the same radial coordinate, R, as those of the paths of approach. In the case of colloidal-size charged latex spheres, however, significant interaction forces due to repulsion between electrical double layers and attraction due to van der Waals forces come into play when sphere surfaces approach to within a distance h = 50 nm (van de Ven and Mason 1976). This results in asymmetric collision trajectories (Takamura et al. 1981), as illustrated in Fig. 2, which can be analyzed and hydrodynamic theory used (van de Ven and Mason 1976 ; Arp and Mason 1977) to show that net interaction forces as small as 10^{-13} N can be detected. In the case of latex spheres in aqueous solutions of simple electrolytes, the interaction forces have been interpreted by applying the DLVO theory of colloid stability (van de Ven and Mason 1976), and thereby values of the Hamaker constant and the retardation parameter of the van der Waals force have been obtained (Takamura et al. 1981)

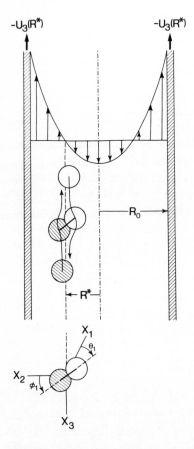

Fig. 1. Two-body collision in Poiseuille flow of rigid spheres forming a transient doublet. The tube, radius R_o, is moved up with a velocity $U_3(R)$ equal but opposite to that of the fluid at the mid-point of the doublet axis, at a radial distance $R = R^*$. Cartesian (X_i) and polar (θ_1, ϕ_1) coordinates are set up at the mid-point of the doublet axis (Takamura et al. 1981).

When the electrical double layer is sufficiently compressed at high electrolyte concentration, or when polymers able to cross-link the surfaces of the interacting particles are present, collisions result in the formation of permanent doublets. In this case, the travelling microtube technique can be used to measure the shear stress required to break up permanent doublets, and hydrodynamic theory used to compute the applied force at break-up (Tha and Goldsmith 1986). We have used this method to measure the hydrodynamic force required to separate doublets of sphered and swollen

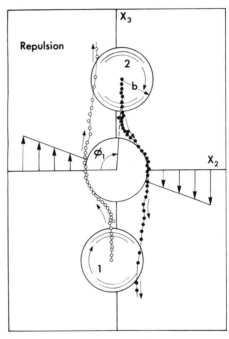

Fig. 2. The projection in the median, X_2X_3-plane of a 200 μm diameter tube of the paths of the centres of two colliding 4-μm polystyrene latex spheres in 50% aqueous glycerol containing 1 mM KCl. The *arrows* along the X_3-axis indicate the velocity profile in the vicinity of the doublet. At the *centre* is the exclusion sphere which cannot be penetrated since the collision occurred in the X_2X_3-plane. Due to double layer repulsion there was a significant increase in the distance of separation of centres along the X_2-axis, from 0.6 μm before, to 3.2 μm after collision (Takamura et al. 1981).

aldehyde-fixed human red cells of antigenic type B cross-linked by the corresponding poly- or monoclonal antibody (Tha et al. 1986 ; Tees et al. 1993). In these experiments the doublets were tracked in a continuously accelerating Poiseuille flow until break-up at a known fluid shear stress.

For reasons explained below, the technique has been extended to Couette flow using a counter-rotating cone-and-plate Rheoscope (Tees et al. 1993).

10.2 The Travelling Microtube

The device shown in Fig. 3 was constructed in the workshop of the Chemistry Department at McGill. It consists of a vertically mounted sliding platform with a built-in microscope stage for mounting the flow tube, and supports a syringe infusion-withdrawal pump or, in the case of gravity flow, infusion and collecting reservoirs. The platform is mounted vertically on support brackets through a set of recirculating ball-bearing gibs ensuring smooth frictionless sliding and preventing lateral motion. Both the platform and the syringe pumps are driven hydraulically by the pistons of slave cylinders connected to master cylinders. The pistons of the latter are driven via micrometer screws by continuously variable speed electronically-controlled DC motor drives. The distance travelled by the platform, a maximum of 5 cm at velocities up to 1 mm s^{-1}, is recorded on a dial indicator. The flow tube assembly consists of a 5 to 6 cm length of precision bore glass tubing having a wall thickness $< 0.3R_o$, rigidly cemented to a stainless steel hypodermic needle with a teflon hub. Tube and needle are laid down the centre of a 5×7.5 cm microscope slide which is held in a jig to provide support. To reduce optical distortion, the tube passes through a chamber filled with fluid of the same refractive index as the particle suspending medium. The jig housing the slide and flow tube is mounted on the stage of the sliding platform and connected at the inflow and outflow to the two reservoirs as shown in Fig. 3.

Observations are made through a monocular tube of a high power microscope, the optic axis being fixed and normal to that of the sliding platform. Both brightfield and interference phase contrast condensers are used with constant or flash illumination by Tungsten or Xenon arc lamps. Particle motions are recorded on 16 mm cine film using a Locam pin registration camera (Red Lake Labs., Santa Clara, CA) at 10 - 500 pictures/s, and on videotape using a model C1000 video camera (Hamamatsu Systems Inc., Waltham, MA) or a CCD video camera with an electronic shutter (Sony Canada, Ltd) to avoid blurring of particles at high velocities.

Fig. 3. The travelling microtube, resting on a vibration-free table, showing the frame **A**, platform **B** and microscope **C**. Particles are tracked in an accelerating Poiseuille flow by gravity feed between an infusion pipette **D** connected to a Harvard pump **E** which infuses silicone oil layered on top of the suspension in the pipette, raising the level of fluid and thereby accelerating the flow rate. Suspension enters the flow tube through PE205 tubing **F**. The flow tube is mounted on a microscope slide **G** which in turn is held by a jig **H**, the whole being attached to the platform by clamps **I**. The suspension flow out into a collection reservoir syringe **J**. Also seen is the slave cylinder **K** driving the platform upwards at a velocity indicated by the dial gauge **L**.

10.3 Interaction Forces in Two-Body Collisions

10.3.1 Theoretical Considerations

In the presence of interparticle forces, $F_{int}(h)$, the trajectories of two equal-sized spheres of radius b in shear flow are given by (van de Ven and Mason 1976 ; Arp and Mason 1977 ; Batchelor and Green 1972) :

$$\frac{dr}{dt} = A(r^*)Gb \sin^2 \theta_1 \sin 2\phi_1 + \frac{C(r^*)F_{int}(h)}{3\pi\eta b} \tag{1}$$

where r is the distance between sphere centres, G is the shear rate and $A(r^*)$, $B(r^*)$ and $C(r^*)$ are known dimensionless functions of r^* (= r/b) which have been documented (van de Ven and Mason 1976 ; Arp and Mason 1977 ; Batchelor and Green 1972). When $F_{int}(h) = 0$ the trajectories of approach and recession of the colliding spheres are symmetrical about the orientation $\phi_1 = 0$ of the doublet axis (Fig. 1), and are defined by two constants, D and E given by the integrated form of Eq. (1) (Arp and Mason 1977 ; Batchelor and Green 1972).

$$x_1 = \pm\frac{1}{2}r^*Df(r^*) \tag{2}$$

$$x_2 = \pm\frac{1}{2}r^*f(r^*)\left[E + g(r^*)\right]^{1/2}$$
(3)

where $g(r^*)$ and $f(r^*)$ are integral functions of r^* for which numerical values have been given (Arp and Mason 1977 ; Batchelor and Green 1972). When $F_{int}(h) \neq 0$, the paths of approach and recession are no longer symmetrical, and D and E become functions of time, increasing when $F_{int}(h) > 0$ (repulsion), and decreasing when $F_{int}(h) < 0$ (attraction).

10.3.2 Determination of the Interaction Force

In principle, $F_{int}(h)$ can be computed from Eq. (1) if one can measure the observed translational velocity, dr/dt. This turns out to be difficult since the trajectory may lie out of the median, X_2X_3-plane, and even for collisions in the median plane it is impossible to measure dr/dt when h <100 nm, just when particle interaction forces begin to be important.

Instead, recourse is had to fitting the experimentally observed trajectories by numerically integrating Eq. (1), and solving for D and E (Takamura et al. 1981). In addition, in the case of the charged latex spheres, by using expressions from the DLVO theory (van de Ven and Mason 1976), it is possible to compute the Hamaker constant, A, for the latex:

$$F_{int}(h) = F_{attr}(h) + F_{rep}(h) \tag{4}$$

where the van der Waals attractive force, $F_{attr}(h)$, is given by:

$$F_{attr}(h) = -\frac{Ab}{12h^2}\left[\frac{1+3.54p}{(1+1.77p)^2}\right] \qquad\qquad p<1 \tag{5A}$$

$$F_{attr}(h) = -\frac{Ab}{12h^2}\left[\frac{0.98}{p} - \frac{0.434}{p^2} + \frac{0.0674}{p^3}\right] \qquad p>1 \tag{5B}$$

and the double layer electrostatic repulsion force, $F_{rep}(h)$, is given by :

$$F_{rep}(h) = 2\pi Kb\varepsilon_0\psi_0^2\left[\frac{e^{-\kappa h}}{1\pm e^{-\kappa h}}\right] \tag{6}$$

Here $p = 2\pi h/\lambda$ is the retardation parameter, λ being the London wavelength, K is the dielectric constant of the suspending medium, ε_0 the permittivity of free space, ψ_0 the surface potential (usually taken to be the measured ζ-potential), and κ the reciprocal Debye double layer thickness. Fig. 4 shows an example of the trajectory plotted as $r^*\sin\theta_1$ (where θ_1 is the initial value of the polar angle) against dimensionless time $t^* = Gt$. The paths of recession were computed for various values of the constant, with a best fit at $A = 6 \times 10^{-21}$ J.

A problem encountered in the study with latex spheres was the observed pronounced repulsion in a few collisions, not accounted for by double layer interaction or by surface roughness. This occurred when the predicted trajectory would have led sphere surfaces to approach to a distance $h < 12$ nm, and may be due to the existence of solvation forces, which in water have been reported to act at distances < 6 nm (Israelachvili and Adams 1978).

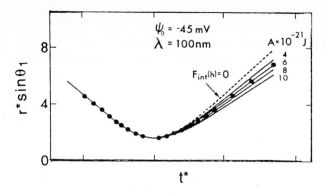

Fig. 4. Plot of the dimensionless projected length, r*sinθ₁, of the doublet axis against dimensionless time, t* = Gt, for the trajectory of a transient doublet of 2.6 μm polystyrene latex spheres in 50% aqueous glycerol containing 10 mM KCl. The points are experimental and show the asymmetry of the trajectory, with the velocity of recession, dr/dt, greater than the velocity of approach. The lines were computed by numerically integrating Eq. (1) with $F_{int}(h)$ obtained by varying the parameter A in Eqs. (5) and (6) with ψ_0 = -45 mV and λ = 100 nm (Takamura et al. 1981).

In the case of sphered and swollen red blood cells, two-body collisions are markedly asymmetric with an apparent repulsion, the trajectory constant E increasing and the minimum h > 50 nm (Goldsmith et al. 1981), i.e. a distance large compared to that at which double layer interactions become important (Capo et al. 1982). The result is no doubt due to the interaction, principally steric, between the glycocalyces of the colliding cells.

10.4 Hydrodynamic Force to Separate Doublets

10.4.1 Theoretical Considerations

Rearrangement of Eq. (1) yields the force equation:

$$\frac{3\pi\eta b}{C(r^*)}\frac{dr}{dt} = \frac{A(r^*)}{C(r^*)}3\pi\eta Gb^2 \sin^2\theta_1 \sin 2\phi_1 + F_{int}'(h) \qquad (7)$$

The term on the left represents the hydrodynamic drag force resisting approach of the particles. The first term on the right is the normal hydrodynamic force acting along the line of centres, $-F_n$, which is maximal at $\theta_1 = \pi/2$ (i.e., rotation in the X_2X_3-plane) at $\phi_1 = -\pi/4$ where it is compressive, and at $\pi/4$ where it is tensile.

If a doublet of rigidly linked non-deformable spheres rotating in a shear

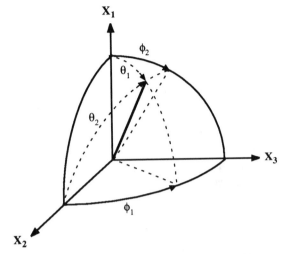

Fig. 5. Orientation of doublet in shear flow : Cartesian and polar coordinates with origin at the mid-point of the doublet axis (*thick line*).

field can be induced to break up, then since $dr/dt = 0$, Eq. (7) reduces to:

$$F_{int}(h) = -F_n = -\frac{A(r^*)}{C(r^*)} 3\pi\eta Gb^2 \sin^2\theta_1 \sin 2\phi_1 \tag{8}$$

We have computed both the normal and the shear force acting on a doublet whose axis may lie in any plane. In terms of the polar and azimuthal angles, θ_1 and ϕ_1, and θ_2 and ϕ_2 shown in Fig. 5, the normal force, F_n, acting along, and shear force, F_s, acting normal to the doublet major axis (thick black line in Fig. 5) are given by (Tha and Goldsmith 1986) :

$$F_n = 19.33\,\eta Gb^2 \sin^2\theta_1 \sin 2\phi_1 \tag{9}$$

$$F_s = 7.02\,\eta Gb^2 \left[(\cos 2\theta_2 \cos\phi_2)^2 + (\cos\theta_2 \sin\phi_2)^2 \right]^{1/2} \tag{10}$$

Using the above expressions and the measured shear stress at break-up of doublets subjected to an accelerating Poiseuille flow, we have calculated the force required to separate two sphered, swollen human red blood cells (SSRC) cross-linked by antibody.

Whether the shear or normal force is responsible for doublet separation is moot. In what follows, the results are given as the maximum normal force acting on a doublet in the last quarter of its rotational orbit before break-up occurs, computed using Eq. (9) (applicable to Poiseuille flow if $b << R_o$) with the experimentally determined values of η, G, b, θ_1, and ϕ_1.

10.4.2 Sphering of Red Cells

Antigenic type B red cells are sphered and swollen in 0.2 M glycerol containing 1.73 mM sodium dodecyl sulphate, fixed in 0.08% glutaraldehyde and washed repeatedly. SSRC are suspended at concentrations of 1.25 or 2.5×10^4 cells/μl in buffered 67 - 76% glycerol at polyclonal anti-B anti-serum concentrations from 0.44 to 2.44 g/l (Tha et al. 1986), and in buffered 46 - 56% sucrose at concentrations of monoclonal anti-blood group B affinity purified IgM or IgA antibody from 0.075 to 0.60 nM (Tees et al. 1993).

10.4.3 Constant Slow Acceleration in Poiseuille Flow

The SSRC suspensions are subjected to an accelerating Poiseuille flow through a 150 μm glass tube by gravity feed between infusion and collecting reservoirs (Tha et al. 1986 ; Tees et al. 1993). The flow rate is uniformly accelerated by pumping silicone oil on top of the cell suspension in the infusion reservoir, resulting in a slow linear increase in centerline velocity ~ 25 μm s^{-2}. Doublets are videotaped while being tracked until break-up, or until they are lost to view at the bottom of the tube, by moving the sliding platform and microscope stage (Fig. 3) upwards at a velocity equal to that of the downward-flowing particles. Measurements of SSRC radius, orientation angles (θ_1 and ϕ_1) and radial distance from the tube axis, R, are made on a video monitor using a superimposed videoposition analyzer calibrated with a micrometer scale. The shear rate used in the force calculation is found from the measured R and doublet linear velocity at break-up, $u_3(R)$ (Tha et al. 1986). Using these values and the glycerol or sucrose viscosity at the recorded temperature, F_n at break-up can be computed from Eq. (9).

Scatter plots of the maximum normal force, F_n, at break-up obtained with polyclonal antiserum and monoclonal IgM antibody are shown in Fig. 6a and b respectively. The mean and standard deviation of the data at each antibody concentration are shown to the right of each column. The plots show that, for both monoclonal and polyclonal antibodies: (i) the normal forces at break-up at different concentrations of antibody exhibit considerable overlap ; (ii) the average force at break-up increases significantly with antibody concentration, and (iii) there is no evidence of clustering at discrete values of F_n (corresponding to the forces required to break different numbers of antigen-antibody bonds).

The first similarity suggests that there is a distribution in the number of antibody-antigen bonds linking the doublets. The average force at break-up, then, increases as the average number of bonds between cells increases. Since bond formation is considered to be a Poisson process (Capo et al. 1982), a large scatter in the number of bonds, and hence F_n, at low bond number is expected. Chi-square goodness of fit and Poisson heterogeneity tests show the data to be consistent with a minimum force of 24 pN for the polyclonal antiserum (Tha et al. 1986) and 20 pN for the monoclonal antibody (Tees et al. 1993).

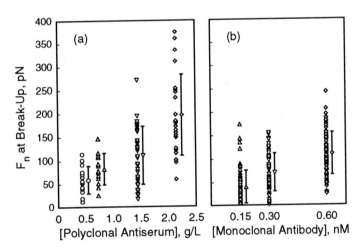

Fig. 6a, b. Scatter plots of normal force at break-up in constant slow acceleration experiments for **a** polyclonal and **b** monoclonal antibody. *Error bars* show one standard deviation from the mean of the data at each concentration.

The lack of clustering at discrete values of F_n for the polyclonal antibody can be attributed to the heterogeneous nature of the antiserum. The lack of clustering at force values corresponding to one bond, two bonds, etc., in the monoclonal preparations, however, is surprising since the experimental error of ± 2 pN is considerably less than the predicted minimum F_n of 20 pN to break up one bond. Furthermore, the above arguments cannot account for the existence of six break-ups at $F_n < 10$ pN (Fig. 6b), forces much lower than the supposed minimum. In any case, the observation of break-ups in steady flow after immediate application of shear, shown below, support the notion that break-up is a time-dependent, stochastic process (Evans et al. 1991). Accordingly, experiments designed to test both the time and force dependence of doublet break-up at a constant shear rate were undertaken using the travelling microtube as well as the Rheoscope.

10.4.4 Rapid Halted Acceleration in Poiseuille Flow

To determine whether doublets subjected to a constant shear stress in each rotational orbit break up after a noticeable time lag, experiments were

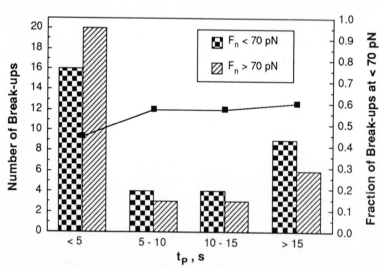

Fig. 7. Time dependence of break-up in Poiseuille flow, rapidly accelerated to a predetermined value of the normal force, F_n, after which acceleration was halted. Histogram of the number of doublets breaking up as a function of time, t_p, after the start of steady flow, divided into four ranges. The data are divided into high ($F_n \geq 70$ pN) and low ($F_n \leq 70$ pN) force ranges. (Tees et al. 1993).

performed in which silicone oil is pumped into the infusion reservoir at a high flow rate for a short time, resulting in a rapid linear acceleration of ~150 μm s^{-2} at the tube axis. After infusing for 5 to 15 s, depending on the desired magnitude of the force (from 30 - 200 pN) and the doublet radial distance from the tube axis, infusion is suddenly halted and the particle tracked under steady flow conditions. To determine the infusion time necessary to attain a given force, the motion of a doublet, once selected, is arrested and measurements of its diameter, ϕ_1- and θ_1-orientation and radial distance made, and along with the known viscosity, entered into a microcomputer to calculate the required shear rate and centreline fluid velocity, $u_3(0)$. The computer is programmed with the known relation between infusion rate and $u_3(0)$ in the tube to signal the observer when the desired value of $u_3(R)$ is reached, at which time the infusion pump is turned off. The time to break-up, t_p, in steady flow is recorded.

Of 280 doublets tracked in the microtube, 24 (9%) broke up during the acceleration phase, and a further 41 (15%) during steady flow. With the exception of 5 doublets, break-up occurred within 30 s of the onset of steady flow. The remaining 76% of doublets did not break up before being lost at the lower end of the tube; of these, 80% were tracked for > 30s. The results show that there is a distribution in time to break up after the onset of steady flow, which is force-dependent. Thus, as shown in Fig. 7, the proportion of the total number of break-ups in the low force range increases from (44%) in the lowest time range (t_p < 5 s), to 60% at t_p > 15 s. The fraction of all doublets tested which broke up also increased with increasing F_n: 8.5% broke up at F_n < 30 pN, whereas 30% broke up at $F_n \geq$ 90 pN.

10.4.5 Couette Flow: Immediate Application of Shear

To avoid uncertainties introduced by the time spent by the doublets during acceleration of the flow in the microtube, we have studied particle break-up in Couette flow using a commercially available transparent, counter-rotating cone and plate Rheoscope system in which the shear stress is almost instantaneously applied (model MR-1, Myrenne Instruments, Freemont CA). Here, the doublets are viewed along the X_2-axis normal to the vorticity axis of the flow field (Fig. 5) and are seen projected onto the X_1X_3-plane, rocking to and fro between the angles ϕ_{2max} (ϕ_1 = 90°) and ϕ_{2min} (ϕ_1 = −90°). Doublets are allowed to form at the lowest shear rate (5.4 s^{-1}), and their rotation recorded on videotape. Flow is then stopped while the SSRC diameter and orientation angles are measured and entered, along with the known viscosity, into a microcomputer to determine the cone and plate rotational velocities required to produce a desired force. The doublet is then observed under high shear until break-up or loss from view. As in Poiseuille flow, F_n is calculated from Eq. 9 using the known medium viscosity, the doublet size and orientation, and shear rate determined from the Rheoscope cone angle and the cone and plate rotation rates.

Of 340 doublets analyzed, 125 (37%) broke up before being lost to view, from 10 - 20 s after flow commenced. As in the rapid halted acceleration runs in the microtube, there was a force-dependent distribution in times to break up. As indicated by Eqs. (9) and (10), in shear flow, the applied force varies periodically, the doublets being subjected to maximum tensile and shear forces twice during each orbit. The distribution of break-ups, plotted as a function of the number of rotations from the onset of shear, therefore yields information on the number of times the particles are exposed to a given force. Using the measured period of rotation, we can compute the fraction of doublets which break up *in a given rotation*, i.e., the fraction of the total number

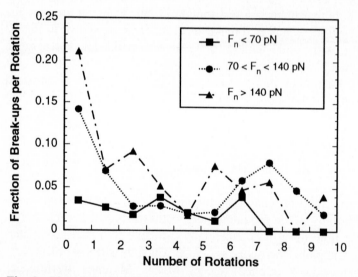

Fig. 8. Plot of the fraction of doublets breaking up per rotation (the fraction of the total number of doublets observed in that rotation which broke up) during the first 10 rotations after the onset of shear, for the high ($F_n \geq 140$ pN), intermediate ($70 \leq F_n \leq 140$ pN) and low ($F_n < 70$ pN) force ranges (Tees et al. 1993).

of doublets observed in that rotation which broke up. Figure 8 shows the fraction of break-ups per rotation for the first ten rotations. With increasing F_n, there is a marked increase in the number of break-ups in the first rotation, followed by an increasingly sharp drop in the second.

10.5   Concluding Remarks

We have described two microrheological techniques for studying interactions between cells in a well-defined shear flow. Both methods have the advantage that the translational and rotational motions of the interacting particles are observed in detail and are recorded for analysis. Moreover, the methodology can be used to study any receptor-ligand interaction provided suitable derivatized erythrocytes or beads can be obtained. The disadvantage of the methods is the fact that only one interaction or one break-up may be studied at a time. The acquisition of data on populations of cells is therefore a lengthy process.

With respect to the study of collisions between cells, data acquisition could be speeded up by having one cell adhere to the wall of the tube, while a suspension of cells flows out of a micropipette with its narrow tip located eccentrically at the tube entry so that collisions continually occur with the fixed cell. Theory for such collisions is available. As regards the study of SSRC break-up, one could conduct a population study in a Rheoscope having a larger sample volume. Here, one would shear a suspension for a given time, and then sample it to measure the numbers of doublets, triplets and quadruplets etc. as a function of shear stress, using flow cytometry. In the case of aldehyde-fixed neutrophils, it has been shown that it is possible to measure the numbers of such aggregates without having to use fluorescent markers, and that doublets, triplets and quadruplets yield separate peaks on the fluorescence intensity histogram (Rochon and Frojmovic 1991).

References

Arp PA, Mason SG (1977) The kinetics of flowing dispersions. VIII. Doublets of rigid spheres (theoretical). J Colloid Interface Sci 61:21-43
Batchelor GK, Green JT (1972) The hydrodynamic interaction of two small freely moving spheres in a linear flow field. J Fluid Mech 56:375-400

Capo C, Garrouste F, Benoliel A-M, Bongrand P, Ryter A, Bell GI (1982) Concanavalin-A-mediated thymocyte agglutination: a model for a quantitative study of cell adhesion. J Cell Sci 56:21-48

Evans EA, Berk D, Leung, A (1991) Detachment of agglutinin-bonded red blood cells. I. Forces to rupture molecular-point attachments. Biophys J 59:838-848

Goldsmith HL, Mason SG (1964) The flow of suspensions through tubes. III. Collisions of small uniform spheres. Proc Roy Soc (London) A 282:569-591

Goldsmith HL, Lichtarge O, Tessier-Lavigne M, Spain S (1981) Some model experiments in hemodynamics. VI. Two-body collisions between blood cells. Biorheology 18:531-555

Israelachvili JN, Adams GE (1978) Measurement of forces between two mica surfaces in aqueous electrolyte solutions in the range 0-100 nm. JCS Faraday Trans I 74:975-1001

Rochon IP, Frojmovic MM (1991) Dynamics of human neutrophil aggregation evaluated by flow cytometry. J Leukoc Biol 50:434-443

Takamura K, Goldsmith HL, Mason SG (1981) The microrheology of colloidal dispersions. XII. Trajectories of orthokinetic pair-collisions of latex spheres in a simple electrolyte. J Colloid Interface Sci 82:175-189

Tees DFJ, Coenen O, Goldsmith HL (1993) Interaction forces between red cells agglutinated by antibody. IV. Time and force dependence of break-up. Biophys J 65:1318-1334

Tha SP, Goldsmith HL (1986) Interaction forces between red cells agglutinated by antibody. I. Theoretical. Biophys J 50:1109-1116

Tha SP, Shuster J, Goldsmith HL (1986) Interaction forces between red cells agglutinated by antibody. II. Measurement of hydrodynamic force of break-up. Biophys J 50:1117-1126

van de Ven TGM, Mason SG (1976) The microrheology of colloidal dispersions. IV. Pairs of interacting spheres in shear flow. J. Colloid Interface Sci 57:505-516

11. Analysis of the Motion of Cells Driven Along an Adhesive Surface by a Laminar Shear Flow

A. Pierres, O. Tissot and P. Bongrand

11.1 Rationale

The experimental approach we shall describe consists of monitoring the motion of cells driven along adhesive surfaces by a hydrodynamic drag expected to be weaker than a single molecular bond. A quantitative analysis of this motion might thus give direct information on bond formation and rupture, since these events must cause detectable changes of cell velocity. Additional use of fluorescence-based methods may enhance the power of this experimental approach by allowing direct visualization of early cell signalling events. It must be emphasized that the use of low shear flow to monitor the interaction between individual cells was pioneered by Goldsmith with the travelling microtube technique (Goldsmith et al. 1981 and Chap. 10 in this book). Note that useful information was obtained with parallel-plate flow chambers used at higher shear rate than described in the present chapter (see e.g. Lawrence and Springer 1991).

11.2 Background

11.2.1 Motion of a Cell Near a Surface in a Laminar Shear Flow

We shall briefly review some basic definitions of fluid mechanics. The interested reader is referred to appropriate textbooks for more details (Sommerfeld 1964 ; Happel and Brenner 1991). As shown on Fig. 1, in a laminar shear flow near a plane wall, the fluid velocity v is everywhere parallel to the plane. Due to the medium viscosity, v is zero on the fluid boundary, and the fluid velocity at any point M is proportional to the

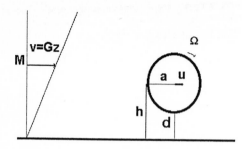

Fig. 1. Laminar shear flow near a planar surface. The undisturbed flow velocity **v** at point **M** is proportional to the distance **z** between **M** and the surface. A freely flowing sphere is endowed with both translational velocity **u** and rotational velocity Ω.

distance z between M and the plane :

$$v = G.z \tag{1}$$

The proportionality coefficient G is the "shear rate" or "velocity gradient". It is expressed in second^{-1}. Now, a cell may be modeled as a sphere of radius a. If the shear rate is low enough, the sphere motion can be predicted by solving creeping flow equations (Goldman et al. 1967). The main results may be summarized as follows :

When the distance z between the sphere center and the plane is much higher than the sphere radius a, the translational velocity is close to Gz, and the rotational velocity Ω is about G/2 (in radian/second). When the ratio z/a decreases, the velocity decreases slowly : thus, the ratio v/Gz is about 0.92 and 0.77 when z/a is 1.54 and 1.13 respectively. The latter value represents a cell-substrate distance of the order of 0.5 µm for a leukocyte of 4 µm radius. The mathematical limit of the sphere translational and rotational velocities is zero when the sphere is in contact with the plane (zero distance). However, this limit is reached very slowly. Thus, the ratio v/Gz is as high as 0.45 when the width d of the gap between the sphere and the plane is 0.3 % of the sphere radius : this corresponds to a value of 12 nm for a sphere of 4 µm diameter. Clearly, it is no longer tenable to model a cell as a smooth sphere for such a small distance.

The relevance of Goldman's equations to actual cells was checked by Tissot et al. (1992), who studied the motion of lymphoid cells along a nonadhesive surface : the mean values of the ratios v/aG and $\Omega/(G/2)$ were

0.86 and 1.18 respectively. It was concluded that the sphere model could be used to describe the cell motion, using for the cell-substrate gap a value close to the length of microvilli (about 1 μm) and adding an extra term to account for the friction force between the cell and the plane surfaces.

A final result of interest is the intensity of the drag force experienced by a sphere of radius a close to the surface. This is given by :

$$F = 32\ \mu\ a^2\ G \tag{2}$$

where μ is the medium viscosity, i.e. about 0.001 Pa.second in water. The force exerted on a standard cell of 4 μm radius is thus :

$$F\ =\ 0.5\ G\ piconewton \tag{3}$$

11.2.2 Strength of a Single Ligand-Receptor Bond

There is no absolute definition of the strength of a single ligand-receptor bond (Bell 1978). Indeed, a bound ligand has nonzero probability to detach at any time, in absence of a disruptive force. However, it seems reasonable to assume that in most cases the mean lifetime of a bond will be decreased by application of a separative force. The bond strength may thus be empirically defined as the force required to decrease the bond lifetime substantially (say, by 50 %). Using this criterion, the strength of a single ligand-receptor bond (e.g. antigen-antibody or lectin-sugar attachment) may be expected to range between 10 and 100 piconewton, as obtained by studying the rupture of doublets of sphered red blood cells agglutinated by minimal amounts of antibodies (Tha et al. 1986) or macroscopically smooth membrane capsules exhibiting molecular point attachment and separated by micromanipulation (Evans et al. 1991).

11.2.3 Expected Translation Velocity of a Cell Bound to a Surface by a Single Bond

The translational velocity v_b of a cell bound to a plane by a single molecular bond may not be zero in presence of a hydrodynamic drag of intensity F. Assuming that the main resistance to motion is generated by the in-plane motion of the cell ligand, we may write:

$$v_b\ =\ F/f \tag{4}$$

where f is the friction coefficient of the plane-bound ligand. If the binding surface is a cell monolayer, the friction coefficient corresponding to a molecule of diffusion coefficient D in the order of $10^{-10}cm^2/s$ would be, following Einstein's formula :

$$f = kT/D = 4 \; 10^{-7} \; newton.second/m \tag{5}$$

Combining equations 3, 4 and 5, a convenient order of magnitude for v_b would be :

$$v_b = 1.25 \; G \; \mu m/s \tag{6}$$

It must be emphasized that the diffusion coefficient of cell membrane molecules may vary between less than $10^{-12}cm^2/s$ and $3 \; 10^{-10}cm^2/s$ or more (see e.g. Bongrand 1994 for a review). When the interaction between rolling cells and substrate-adsorbed molecules is studied, it is advisable to check the mobility of these molecules since wide variations may be expected depending on their structure and mode of attachment to the plane (Micheelis et al. 1980 ; Carpén et al. 1991). Also, if diffusion coefficients are too high, it may not be warranted to neglect the effect of different friction mechanisms on the motion of bound cells.

11.2.4 Minimal Duration of Detectable Cell Arrests

Let Δx be the resolution of cell position measurements. The minimal duration of detectable arrests of a cell of radius a is :

$$\Delta t \sim \Delta x/v \sim \Delta x/aG \tag{7}$$

where the velocity v of rolling cells was approximated as aG (Sect. 11.2.1).

Now, if single molecular attachments are studied, the hydrodynamic drag must be weaker than bond strength F_b. Combining Eqs 2 and 7, we obtain :

$$G \leq F_b/32\mu a^2 \tag{8}$$

$$\Delta t \geq 32\mu a \Delta x/F_b \tag{9}$$

Assuming a value of 10 piconewton for F_b and an accuracy of position measurement of 1 μm, the duration of the shortest detectable arrest of a cell of 4 μm diameter would be of order 0.01 s. Standard video rate would be insufficient to achieve such a sensitivity and special equipment would be required.

11.3 Material and Methods

11.3.1 Flow Chamber

Many authors devised   flow chambers (Hochmuth et al. 1973 ; Baumgartner 1973 ; Doroszewski et al. 1977 ; Sakariassen et al. 1983 ; Forrester and Lackie 1984 ; Mège et al. 1986 ; Lawrence et al. 1987 ; Wattenbarger et al. 1990) and some are described in other chapters of this book. The chamber we used (Tissot et al. 1991, 1992) was suitable for

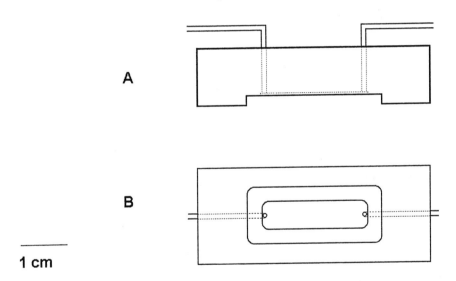

Fig. 2A, B. Parallel-plate flow chamber. Vertical **A** and horizontal **B** views are shown. The chamber body is cut into a perspex block including the fluid inlet and outlet. The floor is a removable coverslip stuck with silicone glue.

both conventional and fluorescence microscopy. As shown in Fig. 2, the upper part was cut into a perspex block and included the fluid inlet and outlet and the chamber cavity. The chamber floor was a glass coverslip that was stuck with silicon glue (Rubson, Brussels, Belgium). Since this glue remained quite soft, it was possible to remove the coverslip after the binding experiment. The chamber was set on the stage of an inverted fluorescence microscope (Olympus, IMT2). If necessary, it was maintained at 37oC with an air curtain.

The flow was generated with a syringe mounted on an electric syringe holder (Razel Scientific, Stamford, CT, supplied by Bioblock, France) equipped with a 2 rpm asynchronous motor. The flow rate could be modified by using different syringes. Another possibility was to change the motor, which was done quite easily. A three-way stopcock allowed manual injection of cell suspensions (about one milliliter) at the onset of each experiment. Here are some points of practical interest :

- In most cases, the chamber floor (10x22 mm^2) was obtained by cutting large (60x22 mm^2) glass coverslips. In some cases, we used plastic culture coverslips (Thermanox ref. 5408, Miles Laboratory, Naperville, IL). However, the optical quality of images was rather poor in this case.
- The glue solvent was fairly acid. This was not a problem when the flow was maintained, but this might impair cell viability under static conditions.
- The wall shear rate may be obtained with standard formulae :

$$G = 6 \ Q/la^2 \qquad\qquad (10)$$

where a and l are the chamber thickness and width respectively and Q is the flow rate. However, this is only an approximation since edge effects are neglected. We found it useful to perform direct checks of the shear rate. For this purpose, the motion of small latex beads (Sigma, St. Louis, MO, 0.8 µm diameter) was studied. The interest of these beads is that they can be seen only if they are in the microscope focus plane (Mège et al. 1985). Thus, the velocity profile was obtained by sequential determination of the mean bead velocity when focusing on planes parallel to the chamber floor and separated by intervals of 5 or 10 micrometers each. These measurements were easily performed with a stopwatch, using a micrometer eyepiece (Tissot et al. 1992).
- When it is desirable to achieve a very low and stable flow rate, it is important to use a syringe holder with an asynchronous motor. More sophisticated devices with computer controlled step motors are unsuitable, since the motion is not smooth enough. A similar problem is encountered with peristaltic pumps.

11.3.2 Microscopic Observation and Image Analysis

Flow apparatus
The apparatus is depicted on Fig. 3. Observations were performed with an Olympus IMT2 inverted microscope bearing a Lhesa 4036 videocamera. The output was connected to a Sopro 600 timer (Soprorep, Marseille) allowing digital display of time values (with 0.1 second accuracy) and small messages. Images were recorded with a standard videotape recorder. Delayed image processing was performed with a PCVision+ Card (Imaging Technology, Bedford, MA) mounted on an IBM-compatible desk computer. This allowed real time digitization of video images with 8-bit accuracy. Digitized images could be processed with a digital-analog converter and displayed on a monitor screen. Bitplane manipulation allowed superposition of a microscopic image and computer-driven cursor as described below. Here are some details concerning this apparatus :

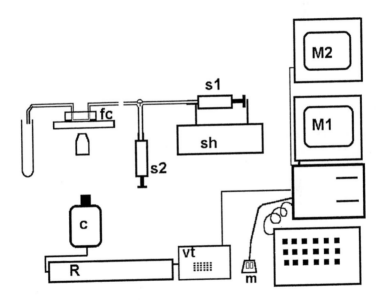

Fig. 3. Flow apparatus. The flow is generated by a syringe (**s1**) mounted on a syringe holder (**sh**). Cell suspensions are introduced with another syringe (**s2**). The flow chamber (**fc**) is set on the stage of an inverted microscope bearing a camera (**C**) connected to a videotape recorder (**R**). The output is sent to a computer through a videotimer (**VT**). Images are subjected to analog-digital and digital-analog conversion and displayed on monitor **M2** with superposion of a cursor driven by the mouse **M**.

- In most cases, observations were performed with a 40x objective, but a 100x immersion lens could be used without any difficulty.
- The Sopro 600 timer is dispensable, since all functions could be performed with the digitizer, but this would require some tedious programming.
- A high sensitivity camera (0.0001-0.001 lux) is not required if fluorescence is not used. However, the interest of this apparatus is to allow minimal sample illumination, which might be useful when the chamber floor is coated with cell monolayers (e. g. endothelial cells).
- Many digitizing cards are commercially available. The PCVision+ card met with the following criteria : first, images recorded with standard VHS material were digitized without any problem. Second, computer access to card memory is possible with minimal interference with the digitization process. Third, image superposition was easily performed by selection of bitplanes for access by the computer and the digitizer.
- Finally, our methodology required no high computer power, and standard XT computers proved quite satisfactory. This point is of lesser importance due to the present availability of low-price high-power desk computers.

Methods
We shall describe two simple procedures allowing rapid acquisition of a high number of cell coordinates to monitor flow properties.

The simplest method is to superimpose on a monitor screen an image of flowing cells and a cursor driven by the computer mouse. Each cell can thus be followed manually with continuous recording of the mouse coordinates together with time. Programming is fairly easy with the PCVision+ card. Cell images are displayed with 7-bit accuracy (using intensity levels 0-127) and the cursor is displayed as a bright line segment (using the 8th bit). When a sampling frequency of 5-10 positions per second is used, the limiting factor for accuracy is the operator rapidity (see below for a quantitative estimate of errors).

Another possible method would be to use a computerized cell-recognition algorithm. As exemplified in Fig. 4, a standard procedure would be to calculate the difference between two sequential images and determine the position of the bright leading front of moving cells. However, this might require fairly high computer power and results might not be safe enough when the images of flowing cells are superposed on an heterogeneous monolayer. Therefore, we used a procedure that took advantage of the observer's ability to recognize moving cells. The basic principle was to follow a suitable cell with the mouse-driven cursor : small (e.g. 32x32 pixel) images were then continuously transferred to host

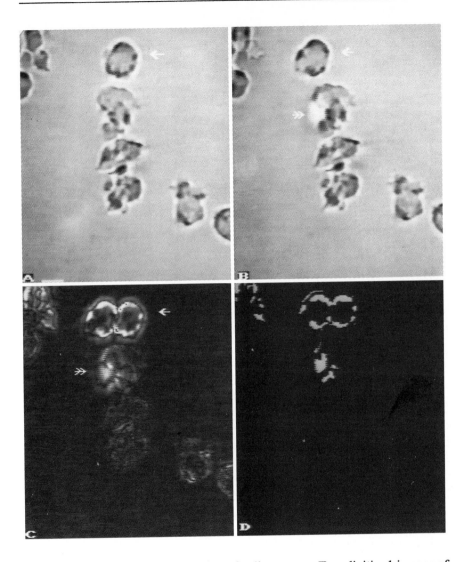

Fig. 4A-D. Automatic determination of cell contours. Two digitized images of flowing cell were obtained   with a time interval of one second (**A** and **B**, horizontal bar is 12 µm).   The pixel per pixel difference between local intensities was calculated (**C**), thus emphasizing moving objects. Minimal image processing (median filtering) and binarization were performed (**D**) to facilitate segmentation

computer memory together with time and cursor position (Fig. 5). A transfer frequency of five to ten images per second was easily achieved, and a simple software allowed delayed recording of sequential cell positions with 1-pixel accuracy.

Fig. 5. Rapid recording of small numerical images. A moving cell was followed manually with a mouse-drive cursor. Small (32x32 pixels) images corresponding to the pointed area were transferred to the computer memory at 0.2 second intervals for delayed analysis. A set of 64 images (included in a 64-kbyte file) is shown.

This work required substantial programming work. The software we developed is available on request. Some details may be of interest. It was found convenient to split programs in two parts. Assembly language was used to write resident programs taking charge of simple tasks requiring optimal speed, such as image transfer. It must be pointed out that the reference manual of the PCVision+ card is very clearly written, and provides all required information for writing programs needed to drive the digitizer. These simple routines were used by a BASIC program that could be modified at will during data processing. It was thus easy to fit the program to any studied system.

An important methodological point is the need to discriminate between cells in contact with the chamber floor and freely flowing ones. This is of obvious importance when the kinetics of bond formation is studied. It is not possible to rely only on microscope focusing, particularly when a low magnification lens is used (the field depth is then too high) and when the adherent surface is not smooth (e.g. when it is coated with a cell monolayer). According to our experience, the most reliable criterion for cell-substrate contact is the frequent occurrence of motion irregularities, due to cell surface protrusions. This results in cell rotation, trajectory curvature, or brief velocity changes. It is advisable to observe the aspect of flowing cells for some time before beginning data acquisition.

10.4. Examples and Data Interpretation

In order to illustrate the potential of the described methodology as well as problems encountered in data interpretation, we shall describe some results obtained by monitoring the motion of lymphoid cells rolling on a glass surface coated with specific antibodies (recognizing CD8 antigen. See Tissot et al. 1993, Pierres et al. 1994b, for a description of the system) or irrelevant antibodies. Other results concerning the interaction between human blood neutrophils and endothelial cell monolayers were described elsewhere (Kaplanski et al. 1993)

10.4.1 Accuracy of Trajectory Determination

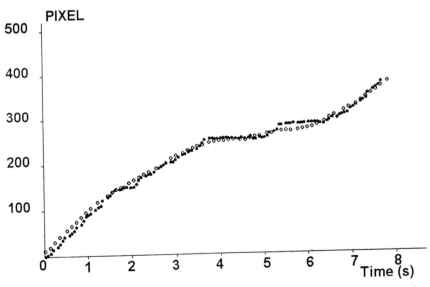

Fig. 6. Typical cell trajectory. Lymphoid cells were driven along an antibody-coated surface with a wall shear rate of 4 s^{-1}. The coordinates of a typical cell were repeatedly determined using the mouse-follow procedure (*full circles*) or the image-transfer method (*open circle*). Positions were plotted as pixel number versus time. A pixel is 0.42 µm wide.

A typical cell trajectory is shown in Fig. 6. Results obtained with both described procedures are shown. Clearly, the observed cell displayed many velocity changes and it is important to know the accuracy of determination of sequential positions to assess the significance of apparent arrests. The following tests were performed :

- A typical rolling cell exhibiting a transient arrest was subjected to tenfold analysis by sequential replay of the passage across the microscope field. The cell was followed with a mouse-driven cursor and positions were recorded twice a second. The mean arrest duration was determined with a standard deviation of 0.21 second. The abscissa of the arrest point was determined with a standard deviation of 1 μm.

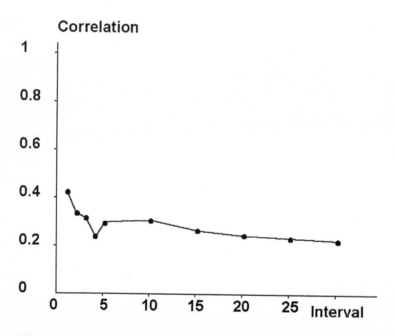

Fig. 7. Correlation between cell velocities measured at sequential intervals. Ten cell trajectories obtained with mouse-follow procedure were used to determine the correlation between instantaneous velocities (determined on a 0.07 second interval) at moments separated by a varying interval (expressed as an integer number of elementary time periods of 0.07 second)

- Ten cells were followed with the cursor procedure using a high sampling frequency (a position recorded every 0.07 second). The correlation between cell displacements at sequential steps was then determined. As shown in Fig. 7, a high correlation between displacements at steps separated by intervals of 0, 0.07 and 0.14 second was found and a plateau value is reached when the time interval is 021 second. Thus, the limit of time resolution is of the order of 0.21 s.

It must be emphasized that the quality of the measurement procedure is critically dependent on the mouse and underlying plane. If inadequate material is used, data acquisition must often be repeated due to inaccurate cell follow.

10.4.2 Definition of Cell Arrests

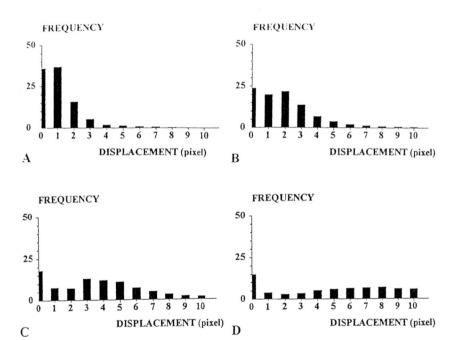

Fig. 8A-D. Distribution of elementary cell displacements. The flow of lymphoid cells driven along antibody-coated surfaces by a shear rate of 4 s[-1] was studied. About 11,000 positions obtained on a sample of 37 cells were used to calculate the distribution of displacements during intervals of **A** : 0.075 second, **B** : 0.15 second, **C** : 0.03 second and **D** : 0.06 second. A bimodal distribution (with peaks supposed to represent free and bound cells) was obtained when long enough time intervals were considered.

A critical point is the definition of cell arrests. This is illustrated in Fig. 8 : the distribution of cell displacements during 1, 2, 4 or 8 sequential steps was determined on a sample of 37 rolling cells. As shown in Fig. 8A, when short time intervals are considered, there is no clearcut difference between moving and arrested cells. On the contrary, when the cell progression over a period of time of 0.3 or 0.6 second (i.e. 4 or 8 elementary steps) is considered (see Fig. 8C and D), the displacement distribution is clearly bimodal, suggesting an objective discrimination between free and bound cells. Since quite different histograms may be obtained with different cell systems, it seems advisable to determine displacement distributions before defining cell arrests.The data shown in Fig. 8 suggested that cells might be considered as bound if they had moved by less than two pixels during a time interval of 0.6 second (i.e. eight elementary steps). It was thus possible to determine the distribution of lifetimes of bound states. if cell

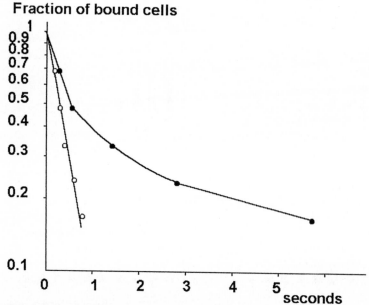

Fig. 9. Two samples of lymphoid cells driven along antibody-coated surfaces were used to determine the distribution of lifetimes of binding events. Two typical curves obtained with samples of 37 and 42 cells are shown. The straight line (*open circles*) is suggestive of arrests being due to single bonds. The curved line (*closed circles*) may be accounted for by an heterogeneity of molecular bonds or formation of additional bonds during cell arrests

arrest is mediated by a single bond, it may be shown with elementary kinetic theory that the fraction $F(t)$ of cells bound at time 0 that will remain attached at time t is given by :

$$F(t) = \exp(-kt) \tag{11}$$

where k is the kinetic constant for bond dissociation. Two typical plots of F drawn on a logarithmic scale are shown in Fig. 9. When a straight line is obtained, it is conceivable that arrest be mediated by a single bond. However, this possibility is by no means proven since the detachment of cells bound to a substrate by multiple links may follow an exponential law after an initial period of time (Pierres et al. 1994a). Therefore, additional experiments are required to obtain more information on the actual number of bonds involved in observed cell stops (e.g. study of the influence of a modification of the density of binding sites on arrest duration). The curved line of Fig. 9 is consistent with the view that in some cases, cell arrest results in formation of additional bonds responsible for a prolonged arrest.

Acknowledgment. This work was supported by a grant from the ARC.

References

Baumgartner HR (1973) The role of blood flow in platelet adhesion, fibrin deposition and formation of mural thrombi. Microvasc Res 5:167-180

Bell GI (1978) Models for the specific adhesion of cell to cells. Science 200:618-627

Bongrand P (1994) Adhesion of cells. In : Lipowsky R, Sackmann E (eds) Handbook of biophysics, vol 1 : Membranes I : structure and conformation. Elsevier, New York. in press

Bongrand P, Golstein P (1983) Reproducible dissociation of cellular aggregates with a wide range of calibrated shear forces : application to cytolytic lymphocyte-target cell conjugates. J. Immunological Methods 58:209-224

Carpén O, Dustin ML, Springer TA, Swafford JA, Beckett LA, Caulfield JP (1991) Mobility and ultrastructure of large granular lymphocytes on lipid bilayers reconstituted with adhesion receptors LFA-1, ICAM-1 and two isoforms of LFA-3. J Cell Biol 115:861-871

Doroszewski J, Skierski J, Przadka L (1977) Interaction of neoplastic cells with glass surface under flow conditions. Exp Cell Res 104 : 335-343

Evans EA, Berk D, Leung A (1991) Detachment of agglutinin-bonded red blood cells. I - Forces to rupture molecular point attachments. Biophys J 59:838-848

Forrester JV, Lackie JM (1984) Adhesion of neutrophil leucocytes under conditions of flow. J Cell Sci 70:93-110

Goldman AJ, Cox RG, Brenner H (1967) Slow viscous motion of a sphere parallel to a plane wall. II - Couette flow. Chem Engn Sci 22:653-660

Goldsmith HL, Lichtarge O, Tessier-Lavigne M, Spain S (1981) Some model experiments in hemodynamics - IV. Two-body collisions between blood cells. Biorheology 18:531-555

Happel J, Brenner H (1991) Low Reynolds number hydrodynamics. Noordhoff Int. Leyden

Hochmuth RM, Mohandas N, Blackshear PL (1973) Measurement of the elastic modulus for red cell membrane using a fluid mechanical technique. Biophys J 13:747-762

Kaplanski G, Farnarier C, Tissot O, Pierres A, Benoliel AM, Alessi MC, Kaplanski S, Bongrand P (1993) Granulocyte-endothelium initial adhesion analysis of transient binding events mediated by E-selectin in a laminar shear flow. Biophys J 64:1922-1933

Lawrence MB, Springer TA (1991) Leukocytes roll on a selectin at physiologic flow rates : distinction from and prerequisite for adhesion through integrins. Cell 65:859-873

Lawrence MB, McIntire LV, Eskin SG (1987) Effect of flow on polymorphonuclear leukocyte/endothelial cell adhesion. Blood 70:1284-1290

Mège JL, Capo C, Benoliel AM, Foa C, Bongrand P (1985) Study of cell deformability by a simple method. J. Immunological Methods 82:3-15

Mège JL, Capo C, Benoliel AM, Bongrand P (1986) Determination of binding strength and kinetics of binding initiation - a model study made on the adhesive properties of P388D1 macrophage-like cells. Cell Biophys 8:141-160

Micheelis I, Absolom DR, van Oss CJ (1980) Diffusion of adsorbed protein within the plane of adsorption. J. Colloid Interface Sci 77:586-587

Pierres A, Benoliel AM, Bongrand P (1994a) Initial steps of cell-substrate adhesion. In : Mowe VC, Guilak F, Tran-Son-Tay R and Hochmuth R (eds) Cell mechanics and cellular engineering. Springer New York, pp145-159

Pierres A, Tissot O, Malissen B, Bongrand P (1994b) Dynamic adhesion of CD8-positive cells to antibody-coated surfaces. J Cell Biol 125:945-953

Sakariassen KS, Arts PAMM, de Groot PG, Houjik WPM, Sixma JJ (1983) A perfusion chamber developed to investigate platelet interaction in flowing blood with human vessel wall cells, their extracellular matrix and purified components. J Lab Clin Med 102 : 522-534

Sommerfeld A (1964) Mechanics of deformable bodies. Academic Press. New York

Tha SP, Shuster J, Goldsmith HL (1986) Interaction forces between red cells agglutinated by antibody. II - Measurement of hydrodynamic force of breakup. Biophys J 50 1117-1126

Tissot O, Foa C, Capo C, Brailly H, Delaage M, Bongrand P (1991) Influence of adhesive bonds and surface rugosity on the interaction between rat thymocytes and flat surfaces under laminar shear flow. J Dispersion Sci and

Technology 12:145-160

Tissot O, Pierres A, Foa C, Delaage M, Bongrand P (1992) Motion of cells sedimenting of a solid surface in a laminar shear flow. Biophys J 58:641-652

Tissot O, Pierres A, Bongrand P (1993) Monitoring the formation of individual cell-surfaced bonds under low shear hydrodynamic flow. Life Sciences Adv : 12:71-82

Wattenbarger MR, Graves DJ, Lauffenburger DA (1990) Specific adhesion of glycophorin liposomes to a lectin surface in shear flow. Biophys J 57:765-777

12 Techniques for Studying Blood Platelet Adhesion

J. J. Sixma and P. G. de Groot

12.1 Introduction

Adhesion of blood platelets to the vessel wall is the first step in the formation of a haemostatic plug or thrombus. Because of this, it is an important and crucial step. Routine techniques for the study of platelet function do not include adhesion studies, partly because no reliable technique is available. Information about adhesive function is sought by indirect means such as aggregation of blood platelets caused by collagen or studies of the von Willebrand factor- Glycoprotein Ib interaction via the use of the obsolete antibiotic ristocetin or the snake venom protein botrocetin.

Adhesion of blood platelets to vessel wall components has a special position among cell adhesion in general, since this adhesion occurs in the normal circulation from flowing blood, i.e. adhesion occurs while platelets are submitted to shear stress. This has as a consequence that studies of platelet adhesion should be performed or supplemented by studies using flow. Special perfusion models have been and are being developed for this purpose. In this chapter we will discuss adhesion tests under static and flow conditions with emphasis on the latter situation.

12.2 Static Adhesion Tests

Static adhesion tests are performed with platelet rich plasma or more often with washed platelets suspended in buffer. Most often, Elisa trays are used that have been coated by incubation with the ligand to be studied, followed by incubation with a suspension of platelets, that are often labelled with either ^{51}Cr or with ^{111}In (Piotrowicz et al. 1988 ; Staatz et al. 1989). In case washing and/or radiolabeling procedures are used, platelets may become activated due to platelet handling and this leads to the activation of the powerful adhesive integrin complex GPIIb-IIIa. In studies in which activation of this complex may influence the test result platelet function, inhibitors such as Prostaglandin (PG)E1 or PGI2 are often used. Although this procedure inhibits activation due to handling, it may also interfere with the actual adhesion process. Adhesion occurs due to gravity, which means that longer incubation times are required for platelet rich plasma than for platelets suspended in a buffer. Usually incubation times varying between 30 min and 1 hour are used and platelet concentrations are often raised to e.g. $1 \times 10^6/\mu l$. After the incubation period the platelet suspension is removed and the Elisa wells are gently rinsed. Quantification of adherent platelets is performed by counting the radioactivity in case of prelabeled platelets or by using a biotinylated platelet-specific monoclonal antibody followed by a peroxidase labelled secondary antibody. After staining for peroxidase activity, the supernatant can be pipetted into wells of a second Elisa tray and read with an Elisa reader. A serious disadvantage of these techniques of studying platelet adhesion is that the result of the evaluation provides an overall figure for the platelets that have remained on the Elisa tray without distinguishing between platelets that have stuck to the primary adhesive coat and platelets that have clumped in aggregates. It is also unknown to what extent platelets have spread on the surface or have remained in a relatively unchanged shape. For this reason, platelet adhesion under static conditions is studied in our laboratory using round glass coverslips which are incubated in 24-well trays with 400 μl PRP or platelet suspension for 30 min - 1 hour at room temperature or at 37°C. After the incubation the coverslips are removed from the wells, rinsed gently with a Pasteur pipet and stained with May Grünwald-Giemsa.

Platelet adhesion is evaluated by a light microscope interfaced with an image analyzer and expressed as percentage surface area covered by platelets. The morphological evaluation allows discrimination between contact and spread platelets and small aggregates. The total number of platelets sticking to the surface cannot be studied in case of large aggregates, but such aggregates are unusual under static conditions.

12.3 Perfusion Studies: Annular Chambers

PERFUSION SYSTEM WITH STEADY FLOW

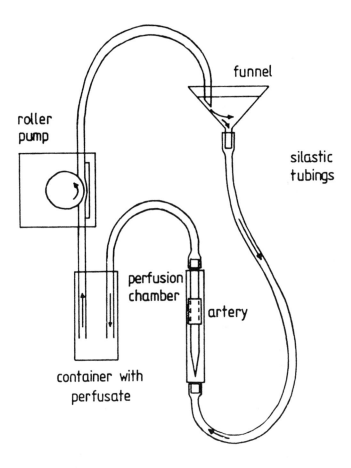

Fig. 1. Drawing of the perfusion system. To avoid the pulsatility generated by the roller pump, a funnel is introduced into the system. The flow rate is determined by the difference in height between the funnel and the chamber.

The first popular perfusion technique that was developed was the annular perfusion technique of Baumgartner and Haudenschild (1972). In this technique, rabbit aorta was everted and mounted on a central rod which fitted into a cylindrical chamber. Anticoagulated whole blood was pumped from a reservoir and circulated through the system at flow rates varying between 10 and 55 ml/min corresponding to wall shear rates between 100 and 2600/sec in laminar flow (Fig. 1). Perfusion times varied from 3 min to 30 min. Evaluation of platelet adhesion occurred by fixation and embedding of the vessel segment followed by evaluation by light microscopy often combined with image analysis. Modifications of the system include the use of a funnel in order to obtain a non-pulsatile flow, the use of human renal and umbilical arteries as vessels and evaluation of platelet adhesion by prelabelling of platelets with either ^{51}Cr or with [111]In. In the latter case platelet aggregate formation should be prevented by the use of e.g. aspirin. Adhesion figures are presented as percentage coverage. Experience has taught us that 40% coverage of the vessel segment corresponds to 40×10^5 platelets/cm^2. Platelet aggregates were originally scored above 5 μm in height, but more recently small aggregates between 2 and 5 μm are also scored separately.

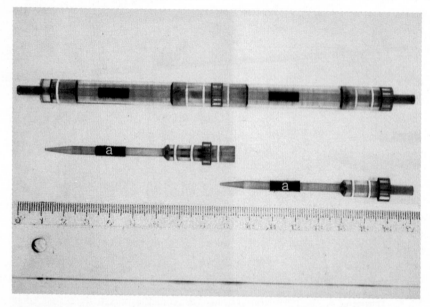

Fig. 2. Example of a double annular perfusion chamber with central rod on which vessel segments (**a**) are mounted.

Modifications of the original chamber have been made for special purposes. These include chambers with a wide annulus for studies at low shear rates and studies with a small annulus for single pass studies with native blood. Recently a double chamber was made to be able to obtain duplicate values in patients with uraemia from whom not much blood can be collected (Zwaginga et al. 1990b ; Fig. 2). Platelet adhesion studies with the annular perfusion chamber are cumbersome and demanding.

Good technique in preparing the vessel segments, in performing the perfusion studies and in evaluating the segments is an absolute necessity. When these conditions are rigorously controlled, good results can be obtained, but the absolute values vary from experiment to experiment, with the vessel segment and the donor blood as main variables. Umbilical arteries, which have become the mainstay in our laboratory, show strong variation between babies and very different results may be obtained with the same segment when blood from different normal donors is used. Part of this latter variation may be due to differences in haematocrit, and levels of fibrinogen (which may inhibit) and von Willebrand factor (which stimulates adhesion), but there are also unknown variables.

As mentioned, the annular chamber has also been used for the study of thrombus formation. For these studies native blood was directly drawn from the antecubital vein through the chamber by using a peristaltic pump. Even when using small diameter chambers, relatively large quantities of blood of 50 ml and more are required for a single run and this is a major limitation. Perfusions with native blood lead also to deposition of fibrin, particularly at low shear rates, and this fibrin is also quantified by light microscopical morphometry (Weiss et al. 1984). A problem may arise when much fibrin is deposited, because platelet thrombi may accumulate on this fibrin network and this may be hard to distinguish from platelet thrombi on the subendothelium. The technique of platelet perfusion studies and the many variables that are involved have been reviewed in a series of authoritative reviews by Baumgartner and his associates, and these reviews should be studied for further information (Baumgartner et al. 1976a, b ; Tschopp and Baumgartner 1975 ; Muggli et al. 1980). Annular perfusion chambers are also useful for studying platelet interaction with biomaterials of catheters, For this purpose the central rod of the chamber is replaced by the catheter.

12.4 Perfusion Studies: Parallel Plate Chambers

Parallel plate or 'flat' perfusion chambers have become popular in recent years to study the adhesion of blood platelets to the extracellular matrix of cultured cells, to special connective tissue molecules coated to a surface and more recently to cryostat cross sections of blood vessels. Several

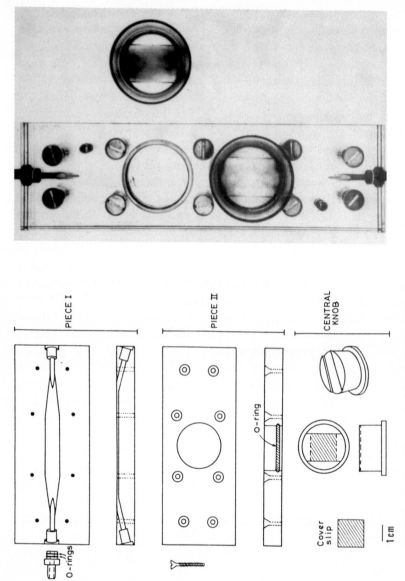

Fig. 3. A Drawing of the perfusion chamber according to Sakariassen mentioned in the text. **B** Photograph of a modified chamber with two plugs.

types of parallel plate chambers have been developed, but our experience has has been with the chamber developed in our laboratory by Sakariassen (Sakariassen et al. 1983 ; Nievelstein et al. 1988) and this is the chamber that we will describe here. The chamber is shown in Fig. 3A and B. The characteristic feature of the chamber is the presence of two plugs on which a depression is present in which a coverslip fits. The coverslips are exposed to the blood flow when the plug is introduced into the chamber. The exposed area of the coverslip has a width of 1.0 cm. Two chambers are in use, one with a height of 1 mm and one with a height of 0.6 mm for different shear rate ranges. The recirculation of blood is similar as used for the annular chamber and shown in Fig. 1.

As adhesive surfaces, we have used extracellular matrices of endothelial cells, vascular smooth muscle cells, and fibroblasts (Sixma et al. 1987 ; Nievelstein et al. 1990 ; Zwaginga et al. 1990a). Several methods for removal of the cultured cells have been tried out among which the use of 2 M urea, 1 mM EDTA and 0.1% Triton X 100. The most reproducible results were obtained, however, with incubation with 0.1 M NH_4OH for 15 min, which removes the cells without affecting the matrix apart from causing a decrease in the von Willebrand factor content. Partial removal of cells, particularly endothelial cells, can be achieved by laying a nitrocellulose strip gently on top of the cells. The cells will stick to this surface and are removed with the nitrocellulose, while the matrix remains. Coating of coverslips with adhesive proteins can be achieved by either incubating the coverslips with the protein solution usually for 1 hour at room temperature or by spraying the protein solution as a fine mist of droplets with a retouching airbrush (Badger model 100, Badger Brush Co, Franklin park, IL, USA). The spraying is performed in such a way that the small droplets dry immediately on the surface, before new droplets are applied. With this procedure a reproducible coating can be obtained. Which way of coating is best for a given protein has to be determined by trial and error. Adhesion studies with fibronectin, collagens, and fibrinogen are routinely performed with sprayed proteins, whereas laminin, thrombospondin, and von Willebrand factor are applied by adsorption.

Evaluation occurs after fixation and staining by May Grünwald-Giemsa by 'en face' light microscopy interfaced with an image analyzer. For evaluation of aggregates, cross sections need to be performed. Thermanox (Miles, Naperville, IL, USA) coverslips are used for this purpose. Adhered platelets are fixed, dehydrated, and embedded in epon. The epon is then removed from the coverslip by rapid immersion in liquid nitrogen which causes the epon to break away from the coverslip, because of the difference in coefficients of expansion. Adhered platelets can also be studied by scanning EM, transmission EM of whole mounts, or by cryostat cross sectioning followed by immuno-electronmicroscopy with radiolabeled gold. For the latter applications, coverslips covered with a melamine foil are used (Westphal et al. 1988). This foil is inert towards platelets, sticks well to the coverslip during the perfusion studies and can easily be moved from the coverslip together with the covering adhering cells after fixation. An interesting and useful application of the parallel plate perfusion chamber is to mount cryostat cross sections of blood vessels on a coverslip in order to be able to study the interaction of platelets with the various layers of the vessel wall. Such an approach is particularly useful for the study of the thrombogenicity of an atherosclerotic lesion. For these studies May Grünwald-Giemsa staining cannot be used, but a biotinylated monoclonal antibody directed against glycoprotein Ib on the platelet membrane, followed by a avidin-peroxidase complex allows visualisation of platelets against the background of non-stained tissue.

Special perfusion chambers have recently been devised in which an obstruction is placed in order to mimic a vessel stenosis (Lassila et al. 1990). Such systems are very useful for mimicking conditions occurring in stenosed vessels and for studying platelet activation occurring at high shear stresses. The perfusion systems currently in use have as a disadvantage that they require relatively large amounts of blood (10-15 ml) for a single run. It has been our practice to use duplicate runs, each with two coverslips giving four values for a single condition. This means that some 15 different conditions can be studied with one unit of blood. This is also more or less the upper limit in number of perfusions that can be performed, when as is usually the case, perfusion times of 5 minutes are used, because times of longer than 5 hours between blood collection and performance of the perfusion studies are undesirable.

A disadvantage of the recirculating perfusion system is that platelet activation may occur during recirculation. It is routine in our laboratory to check for this by studying the presence of small platelet aggregates in the blood at the end of the perfusion run, by comparing the platelet count after paraformaldehyde fixation with the count after EDTA. Activation by recirculation can be prevented by using a single pass system, but these systems tend to use more blood. Small chambers that will not have this problem are currently being developed in our laboratory.

References

Baumgartner HR, Haudenschild C (1972) Adhesion of platelets to subendothelium. Ann N Y Acad Sci 201:22-36

Baumgartner HR, Muggli R (1976a) Adhesion and aggregation: morphologic demonstration, quantitation in vivo and in vitro. In: Gordon JL (ed) Platelets in Biology and Pathology. North Holland, Amsterdam, 23-60

Baumgartner HR, Muggli R, Tschopp TB, Turitto VT (1976b) Platelet adhesion, release and aggregation in flowing blood: effects of surface properties and platelet function. Thromb Haemost 35:124-138

Lassila R, Badimon JJ, Vallabhajosula S, Badimon L (1990) Dynamic monitoring of platelet deposition on severely damaged vessel wall in flowing blood. Effects of different stenoses on thrombus growth. Arteriosclerosis 10:306-315.

Muggli R, Baumgartner HR, Tschopp TB, Keller H (1980) Automated microdensitometry and protein assay as a measure for platelet adhesion and aggregation on collagen-coated slides under controlled flow conditions. J Lab Clin Med 95:195-207

Nievelstein PFEM, d'Alessio PA, Sixma JJ (1988) Fibronectin in platelet adhesion to human collagen types I and III. Use of nonfibrillar and fibrillar collagen in flowing blood studies. Arteriosclerosis 8:200-206

Nievelstein PFEM, De Groot PhG, d'Alessio PA, Heijnen HFG, Orlando E, Sixma JJ (1990) Platelet adhesion to vascular cells. The role of exogenous von Willebrand factor in platelet adhesion. Arteriosclerosis 10:462-469

Piotrowicz RS, Orchekowski RP, Nugent DJ, Yamada KY, Kunicki TJ (1988) Glycoprotein Ic-IIa functions as an activation independent fibronectin receptor on human platelets. J Cell Biol 106:1359-1364

Sakariassen KS, Aarts PAMM, De Groot PhG, Houdijk WPM, Sixma JJ (1983) A perfusion chamber developed to investigate platelet interaction in flowing blood with human vessel wall cells, their extracellular matrix and purified components. J Lab Clin Med 102:522-535

Sixma JJ, Nievelstein PFEM, Zwaginga JJ, De Groot PhG (1987) Adhesion of blood platelets to the extracellular matrix of cultured human endothelial cells. Ann N Y Acad Sci 516:39-51

Staatz WD, Rajpara SM, Wayner EA, Carter WG, Santoro SA (1989) The membrane glycoprotein Ia-IIa (VLA-2) complex mediates the Mg^{++}-dependent adhesion of platelets to collagen. J Cell Biol 108:1917-1924

Tschopp TB, Baumgartner HR (1975) Physiological experiments in hemostasis and thrombosis. Brit J Haematol 31:221-229

Weiss HJ, Turitto VT, Vicic WJ, Baumgartner HR (1984) Fibrin formation, fibrinopeptide release and platelet thrombus dimension on subendothelium exposed to flowing native blood: greater in factor XII and XI than in factor VIII and IX deficiency. Blood 63:1004-1015

Westphal C, Horler H, Pentz S, Frosch D (1988) A new method for cell culture on an electron-transparent melamine foil suitable for successive LM, TEM and SEM studies of whole cells. J Microsc 150:225-231

Zwaginga JJ, de Boer HC, IJsseldijk MJW, et al (1990a) Thrombogenicity of vascular cells. Comparison between endothelial cells isolated from different sources and smooth muscle cells and fibroblasts. Arteriosclerosis 10:437-448

Zwaginga JJ, IJsseldijk MJW, Beeser-Visser N, De Groot PhG, Vos J, Sixma JJ. (1990b) High von Willebrand factor concentration compensates a relative adhesion defect in uremic blood. Blood 75:1498-1507

13. Interference Reflection Microscopy and Related Microscopies and Cell Adhesion

A. S. G. Curtis

Interference reflection microscopy (IRM) produces images of the contact zone between a live cell and a planar substrate. Thus it provides a means of detecting the nature and size of the adhesive contact that a cell may form on an appropriate surface. The image arises from interference between the reflection from the plasmalemmal interface and that from the interface with the substratum, see Fig. 1 below. In some situations these two interfaces are one, namely when there is no gap between the cell and substrate and the adhesive contact is a smooth molecular one. In others there is an extensive, perhaps even continuous gap of some 5 nm to 30 nm between the two interfaces. At the moment only this technique and the closely related techniques of Fluorescence Energy Transfer microscopy and of Total Internal Reflection Fluorescence Microscopy allow visualisation and measurement of the features of the contact zone in living material. IRM led to the discovery of the focal contact (Izzard and Lochner 1976). The technique has been carelessly renamed by some 'reflection interference contrast microscopy'; since all useful types of microscopy develop contrast the use of the word contrast is redundant. There are reasons to suspect that preparation methods for electron microscopy lead to much possible artefact in the contact zone : these possible problems are discussed later in this chapter. IRM and TIRFM avoid such problems.

13.1 Interference Reflection Microscopy (IRM)

13.1.1 Image Formation

The origin of the interference between the two interfaces (plasmalemma-gap and gap-substratum) is outlined in the ray path diagram below. The intensity and order of the reflected intensity is determined by the refractive indices of the three media involved. The conditions for image formation

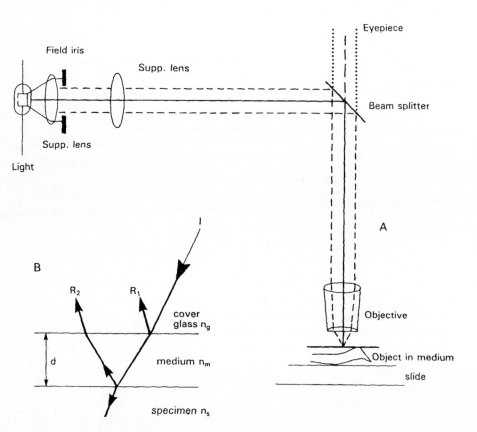

Fig. 1. Raypath for IRM.

will be explored and it will be shown that the contrast ranges and intensities are consistent with small ranges of refractive indices for the plasmalemma, the gap between cell and substratum and the substratum as well as for a limited range of gap thicknesses. In general the technique shows that for many cell types the regions of closest approach of cell and substratum are confined to limited regions, called the focal contacts. Much of the cell substratum contact has a relatively wide gap between the cell and the substratum. The technique is open to artefact and to misleading results : these hazards will be described.

Any good epi-illumination microscope can be converted into an Interference Reflection microscope by replacing one of the dichroic filter blocks with a neutral beam splitter and removing any other excitation or barrier filter.

The major source of degradation in the IRM image comes from reflections from irrelevant surfaces. These fall into two main groups. The first is from refractive index discontinuities, such as organelles inside the cell and on the other (far) side of the cell. If the system has a major refractive index boundary, such as a slide, the interface with this may contribute out-of-focus reflection components to the image : these form the second group of image degradation. Use of high NA objectives with a small depth of focus and slides or holders with an appreciable depth of medium on the far side of the cell will help to reduce these contributions to the image. Deconvolution methods have been used by me to remove this source of image degradation. The second source of image degradation arises from sources of scattered light in the incident ray paths. This problem can be cured by adding a field iris close to the lamp and by painting internal wall surfaces of the objectives with optical matt black paint.

The objectives used should be, as I wrote above, of high numerical aperture and NA values of 0.9 and upwards are the only effective NA values for use with this type of system when very narrow zones such as the contact region are to be studied without image contributions from the other side of the cell or from internal components of the cell. It may be useful to measure the effective NAs of the system with an apertometer. An objective fitted with an iris may be useful to remove scattered incident light. The system should be used both with white light and with monochromatic illumination. A set of neutral density filters which can be placed in the incident light path will probably be helpful to the viewer, especially if the system is run at low light level and visualised through a sensitive video camera. I recommend low light level operation since this diminishes the risk of photodamage to the cells. The commercial IRM system produced at one time by Leitz contains undesirable features which though enhancing contrast introduce reflection contributions from parts of the cell remote from the contact zone.

In principle a confocal microscope operating in the incident light mode should be capable of producing IRM images. In fact, the few such images that have been published, see Vesely et al. (1992), have been rather disappointing in quality and contain surprisingly greater depth of focus than would be expected.

13.1.2 Image Interpretation

In principle, the relationships between incident illuminations, interference intensity give a measure of the gap thickness for monochromatic illumination if the refractive indices of the systems are known. Thus the method can yield quantitative measures of gap thickness and contours can

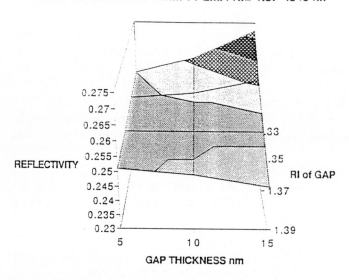

IMAGE REFLECTIVITY : IRM : PLMA RI= 1.37 :546 nʍ

Fig. 2. Relationships between gap thickness and refractive index of gap for plasmalemmal refractive index 1.37. Note that as gap refractive index rises reflectivity in per cent passes through a zero gradient contrast line (for gap refractive index equalling plasmalemmal) to a situation where the reflectivity drops with increasing gap thickness.

be drawn of these thicknesses. Borrowing a term from geology these contours should be known as isopachytes. The earliest interprepations were made by Vasicek (1960) for non-microscopic situations. Subsequently these were repeated with respect to image formation of biological specimens.

Exploration of Vasicek's relationship reveals that the image intensity is very sensitive to the refractive indices chosen for the system. The substratum refractive index can be defined by the use of particular glasses,

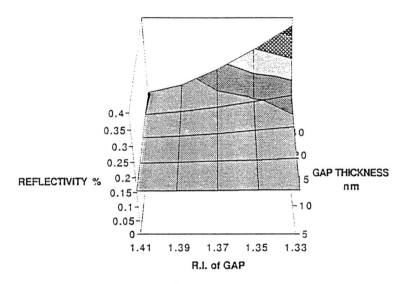

Fig. 3. If the plasmalemmal refractive index were as high as 1.4 the reflectivities would vary more steeply with gap thickness. In both this and in Fig. 2 the reflectivity would drop for any gap thickness as the refractive index of the gap material rises.

or quartz or high refractive index materials such asd lithium niobate. The refractive index of any intervening medium must be that of salty water or perhaps slightly higher if appreciable amounts of organic material are in it. Some of the relationship between gap separation and the various refractive indices are shown in Figs. 2-3. The background reflection intensity is plotted as a horizontal horizon. When the sheet for the intensity for IRM is plotted it can be seen that for some conditions the IRM intensity lies above that for the background ie an image appears brighter than background. Exploration of these conditions shows that the refractive index of the plasmalemma must lie in the very narrow range 1. 34 to 1. 37 if the image is to be darker than background. If it lay outside these limits the image of the cell, if seen at all, should be brighter than background for all reasonable separations.

13.1.3 Summary

The method can be compared with other methods of visualising cell contacts on living cells, i.e. Total Internal Reflection Fluorescence Microscopy and Fluorescence Energy Transfer methods. FET methods have the potential for revealing those situations where two molecules lie within 5-6 nm of each other. Thus cells adherent to a fluorescent substratum might be close enough at places for energy transfer to take place between excited molecules in the substratum and fluorochromes attached to various plasmalemmal molecules. Preliminary work with fluorescein-labelled fibronectin substrates and sulphorhodamine labelled concanavalin A attached to cells failed to reveal any transfer. This preliminary evidence tends to suggest that surface and cell lie a small distance apart in many situations where the cells are clearly in adhesion. Recently surface plasmon resonance methods have been suggested as a method of measuring the thickness of the contact region by Dr. H.Morgan and myself.

13.2 Further Development of the IRM System

The technique can be developed in a number of ways to provide more information on cell adhesion.

In the first, the IRM image is observed while the cell is treated with various reagents that affect adhesion, e.g. protein kinase inhibitors. There are considerable advantages in using a low light level video camera to observe such effects because the use of a low light level reduces the possibility of damage to the cells by photolytic events. The conventional methods of IRM microscopy require a very high light flux since only about 0.2% of the light is refected to the observer.

Second, IR microscopy allows very early events in the penetration of invasive cells through cellular layers to be detected, see work by Verschueren et al. (1991). For example, we have visualised the penetration of lymphocytes through endothelia, and very early events when penetration has only just started can be observed.

Third, IRM can be used to separate events of cell activation from those of adhesion. Molecules which might act on cell adhesion either indirectly or directly can be tested for their mode of action using IR. The molecules are attached to high refractive index beads which can be detected by IRM through several micrometers of cytoplasm. If the beads are located on both adhesive and non-adhesive sides of the cell, no clear distinction can be made between the two theories. But if adhesion is stimulated by beads located only on the non-adhesive side of the cell, an activation event through the cell must occur. The positions of the beads are determined by IRM. When Fibronectin-bearing beads are bound in suspension to fibroblasts the cells will then adhere to surfaces such as adsorbed haemoglobin which they otherwise would have found to be non-adhesive. The beads are found to be on the non-contact side of the cell. This method was first described by Curtis et al. (1992) and recently has been extended to a wide range of molecules of interest applied to the cells by the bead method.

Fourthly, by using very small particles of high refractive index (e.g. gold beads) which may penetrate the gap between cell and substratum, permeation rates can be related to gap thickness and thus gap dimensions and viscosity measured. In addition by a judiciou exploration of the effects both of permeant refractive index and substratum refractive index an unequivocal measurement both of plasmalemmal refractive index and of gap thickness can be obtained.

14.3 Other Methods of Examining the Gap

13.3.1 TIRFM.
See Chapter 14, by Gingell, in this book.

13.3.2 Fluorescence Energy Transfer
This depends upon the Forster effect, whereby energy transfer can occur in two fluorescing centres that are in close proximity. In principle, two molecules that are in very close proximity should show energy transfer if they carry the appropriate fluorochromes when the primary fluorochrome centre is excited. Fluorescein and rhodamine could be appropriate groupings for energy transfer if they lie within 5 nm of each other. Since the intensity of energy transfer declines as the inverse power to the six with distance the method is in principle available for measurements in the range 1 to 5 nm. Thus the method should allow detection of whether an adhesive substratum and an adhering plasmalemma lie at 1 or 5 nm separation. If the technique could be developed, it would allow investigation of those contacts for which IRM or TIRFM give ambiguous results. To date, results from my own laboratory have failed to show energy transfer

between substratum and plasmalemma though model system have worked well.

13.4 Conclusion

IRM is not capable of making a wholly clear distinction between molecular apposition of a cell to a substratum but suggests that extensive areas of low refractive index (close to that of water) are present between cell and substratum. Preliminary FET results lead to the same conclusion and suggest that even focal contacts have a gap of 10-20 nm in the adhesion. Taking these lines of argument and the results on adhesive molecules being activators of cell adhesion it is plausible to return to earlier suggestions that cell contacts are often ones not only established but also maintained by long-range interactions. This, in turn,   raises questions about the methods by which contacts are stabilised from the effecst of proteolysis or oxidation.

References

Curtis ASG (1964) The mechanism of adhesion of cells to glass. J Cell Biol 20:199-215

Curtis ASG (1989) Interference reflection microscopy. In: Lacey A (ed) Light microscopy in biology : a practical approach, IRL Press, Oxford, 45-49   (part of Chap. 2)

Curtis ASG, McGrath M, Gasmi L (1992)   Localised application of an activating signal to a cell: experimental use of fibronectin bound to beads and the implications for mechanisms of adhesion. J Cell Sci 101:427-436

Gingell D, Todd I (1979) Interference reflection microscopy: a quantitative theory for image interpretation and its application to cell-substratum separation measurement. Biophys J 26: 507-526

Izzard CS, Lochner RL (1976) Cell-to-substrate contacts in living fibroblasts: an interference reflexion study with an evaluation of the technique . J Cell Sci 21:129-159

Vasicek A (1960) Optics of thin films. North Holland, Amsterdam

Vesely P, Jones SJ, Boyde A. (1992) Video-rate confocal reflection microscopy of neoplastic cells : rate of intracellular movement and peripheral motility characteristic of neoplastic cells line (RSK4) with high degree of growth independence in vitro. Scanning 15: 43-47

Verschueren H (1985) Interference reflection microscopy in cell biology: Methodcology and applications. J Cell Sci 75:279-301

Verschueren HP, De Baetselier, Bereiter-Hahn J (1991) Dynamic morphology of metastatic mouse T-lymphoma cells invading through monolayers of 10T1/2 cells. Cell motil. cytoskel 20:203-214

14 Advances towards the Measurement of Cell Contacts with Surfaces at Near Nanometer Vertical Resolution by Means of Total Internal Reflection Fluorescence

D. Gingell

14.1 Introduction

Total internal reflection fluorescence (TIRF) provides the most beautiful, the clearest and most easily interpretable images of the contacts between living cells and optical surfaces. The only other technique that can approach it is the far more commonly used method of interference reflection microscopy (IRM), whose appealing simplicity has led to its widespread adoption by cell biologists, by whom it is commonly used with unflinching disregard for the subtleties of image formation which can spawn gross qualitative errors of interpretation. Not that TIRF can be used without thought, but it is superior in that its careful application resolves the difficulties that can bog IRM down in a virtually insoluble morass of complications. TIRF provides a very sharp razor. It is capable of vertical resolution of around 20Å, which is comparable to scanning electron microscopy. Lateral resolution is probably ~ 0.1 μm, typical of light microscopy, though there is currently no theory for assessing it rigorously.

Although TIRF has long been used for studying interfacial chemical processes, in particular protein adsorption (Harrick 1967) its introduction to cell biology is relatively recent (Axelrod 1981 ; review Axelrod et al. 1984). The latter used the membrane phospholipid analogue DiI to label fibroblasts and was able to demonstrate fluorescence at cell-glass contacts. Weis et al. (1982) obtained TIRF images of leucocytes attached to a supported bilayer membrane. Gingell et al. (1985) introduced the variant termed TIRAF, in which the *aqueous* medium contained a low molecular weight fluoresceinated dextran. Where cells make adhesions to glass, a dark image is seen against a brightly fluorescent background. Images in reversed contrast were described by Lanni et al. (1985), who introduced fluorescein derivatives into the cytoplasm of fibroblasts and examined the cells by TIRF. These methods will be discussed in detail below.

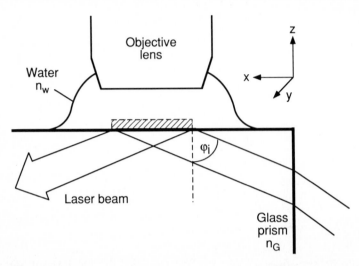

Fig. 1. Basic set up for TIRF. The evanescent wave zone (*hatched*) is shown much thickened. In practice a glass slide is optically mated to the top of the prism with lens oil to provide xy movement of attached cells by means of a modified computer controlled stage drive.

Figure 1 shows the basic set up for microscopic examination of cells by TIRAF. A water immersion objective is focused on the glass surface of a rectangular prism, at the region where a focused laser beam is internally reflected within the prism. The angle of incidence φ_i must exceed the critical angle,

$$\varphi_i > \varphi_{crit} = \sin^{-1}(n_w / n_g) \tag{1}$$

where n_w and n_g are the refractive indices of the aqueous medium and the glass prism respectively. Under these conditions there is an evanescent wave set up in the water, very close to the glass. It exists only within the area where the laser beam is incident. The evanescent wave travels across the interface (*x* direction shown) as a sinusoidal disturbance, but differs from ordinary light in that its energy falls exponentially in the z-direction, perpendicular to the interface. It is convenient to define a distance $\bar{z}$, called the penetration depth, at which level the evanescent wave energy has fallen to $1/e$ of its value at the glass surface ($z = 0$).

$$\bar{z} = \frac{\lambda_{vac}}{4\pi(n_g^2 \sin^2 \varphi_i - n_w^2)^{1/2}} \qquad (2)$$

Where λ_{vac} represents the vacuum wavelength of the incident light.
Thus the energy at a depth z is

$$E^2(z) = E_0^2 e^{-z/\bar{z}} \qquad (3)$$

where E^2 represents the squared amplitude.

This is true whatever the state of polarization of the laser beam, but the waveform of the evanescent waves depend upon the polarization of the laser beam. I shall discuss only s-polarized light (the direction of the electric field is parallel to the top right hand edge of the prism in the diagram) because the equations, especially for the multilayer models (see later), are less clumsy. The case of p-polarization is treated by Heavens and Gingell (1991).

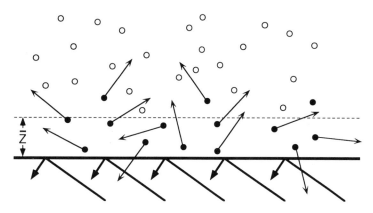

Fig. 2. Side view of the evanescent zone. Fluorescein molecules (**O**) diffusing into the evanescent zone (characteristic thickness $\bar{z}$) absorb energy from the laser beam and emit fluorescence (l) in all directions

In a perfect situation, the evanescent wave remains localized at the interface, no laser photons leave it, and no hint of the electromagnetic events at the surface are discernible by eye. In reality, small imperfections on the glass surface cause a certain amount of the incident laser radiation to be scattered in all directions and some can be seen if a microscope is focused on the interface. But if the water on the prism is replaced by an aqueous solution of a fluorescent dye, which can be excited by the 488 nm emission of an argon ion laser, bright fluorescence is emitted. As shown in

Fig. 2 the molecules of fluorescein which diffuse into the evanescent zone

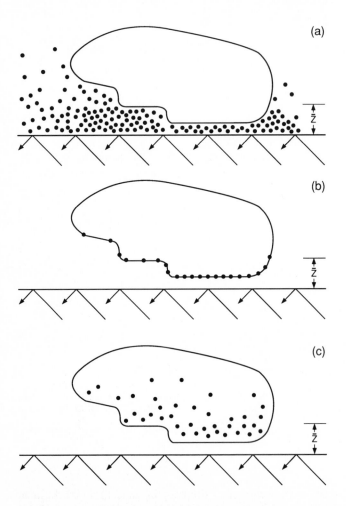

fig. 3a-c. A schematic cell in side view attached to a TIR surface. **a** TIRAF in the presence of aqueous fluorescein. The density of the dots, representing molecules emitting fluorescence, decreases exponentially in the z direction. Fluorescence is thus related to the size of the aqueous gap, asymptoting to the (maximum) background value. **b** The membrane carries a fluorescent dye such as DiI (or a fluorescent antibody). The strength of emission is related to the distance of the membrane from the TIR surface. **c** The converse of **a** in which the cytoplasm is labelled with a fluorescent dye.

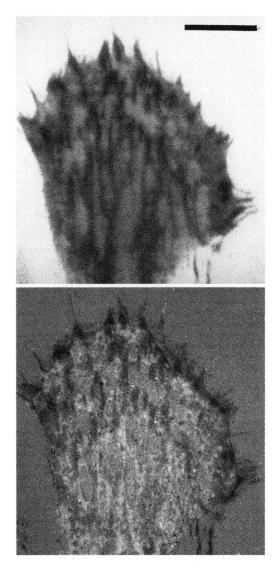

Fig. 4. (*Upper*) TIRAF image of chick heart fibroblast in presence of 4000 M.W. dextran labelled with fluorescein (FD4). Comparison with IRM image (*lower*) shows that the TIRAF image is not affected by cytoplasmic granularity. Focal contacts (*black*) clearly correspond in the two images, but broad close contacts are far easier to distinguish under TIRAF. (Methods as in Gingell et al. 1985). Scale bar=10μm.

become excited and emit fluorescence. Molecules beyond the range of the evanescent wave do not become excited. This provides the basis of TIRAF (Fig. 3a). If a cell or a sphere which is impenetrable to the dye is close to the surface, it displaces fluorescein. Its image in the microscope appears as a dark spot, brightening radially until the maximally bright background field value is reached. The fluorescence at any particular distance within the range of the evanescent wave (roughly $3\bar{z}$) can be measured. When this is divided by the background, the ratio obtained can be used to uniquely characterize the distance if the appropriate refractive indices are known. It is usually the case that the cell thickness much exceeds $3\bar{z}$ so that no fluorescence excitation occurs on the far side. Figure 4 shows a fibroblast seen by TIRAF with an IRM image for comparison. Where the

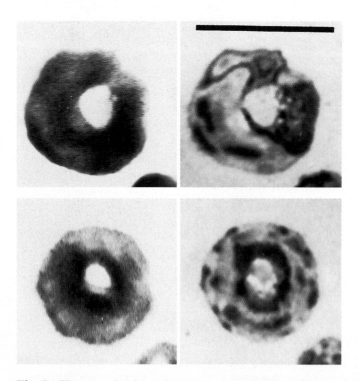

Fig. 5. Human platelets spread on microscope slide n=1.525 in plasma containing FD10   (0.1 mg ml^{-1}) *Left*, TIRAF ($\bar{z}$=110nm; λ =488nm); *right* IRM. Platelets have a central cup-shaped depression (bright in TIRAF). See text for details. Scale bar=10μm.

cell is closest to the glass, the image is dark; where it is further away, the image is less dark. The background is bright. Comparable images of human platelets are given in Fig 5. The central cup-shaped depression is evident and around it the cells are more or less closely adherent to the glass. These platelets exhibit signs of extreme thinning which generates dark patches in the IRM images. In some of these regions the evanescent wave can reach through the cytoplasm and cause fluorescence on the far

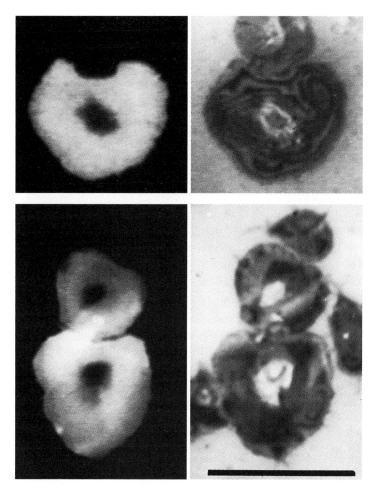

Fig. 6. Human platelets labelled with $C_{18}DiI$. *Left,* TIR; *right* IRM. Upper cell illuminated at reduced penetration $\bar{z} = 61nm$. Lower cell $\bar{z} = 110$ nm, the uniform image beyond the central cup indicates uniform contact. Complex interference fringes in the IRM image (INA ~ 1.2) show that the free surface of the cell is undulatory. Scale bar=10μm.

side, noticeably lightening the otherwise dark image. This phenomenon is discussed in more detail below.

There are two further ways in which distance information can be obtained. Figure 3b shows a cell with fluorochrome molecules associated with its membrane. The strength of excitation, and hence fluorescence emission from the more distant parts of the membrane, is reduced because the evanescent wave energy falls in the z direction from the glass surface. Such images have been obtained using the hydrophobic carbocyanine dye DiI by Truskey et al. (1992) on fixed endothelial cells using a high value of $\overline{z}$ equal to 123nm. Figure 6 shows labelled platelets spread on a glass surface. At moderate $\overline{z}$ (upper cell) the TIR image reports only the topography of the lower membrane. Around the central cup is a zone of uniform apposition. The lower cell is illuminated at $\overline{z}$ = 110nm and the wave is seen to penetrate the cell and show up bright crystals of DiI on the free surface. However, although these images are most attractive, the calculation of absolute distances is difficult (see Truskey et al. 1992 for a method of obtaining relative distances), as there is no background value to relate to the fluorescence of the object, though distances can be computed from fluorescence measurements made at several angles of incidence. A further disadvantage of this method is that the label is generally at an unknown concentration in the membrane and the dimethylsulphoxide commonly used to get it into aqueous medium may perturb cell function. The third approach, introduced by Lani et al. (1985), is to label the cytoplasm (Fig. 3c). Membrane permeant dyes may be used to achieve this. We have used the probe carboxy SNARF-AM, a dye used for fluorometric measurement of intracellular pH, which has the useful properties of membrane penetration (due to the acetomethoxy group) and inability to fluoresce until the carboxyl group is hydrolysed by intracellular esterases (Shahbazi et al. in preparation). But here, too, arises the problem that there is no background to measure, and in addition, dye can become sequestered in intracellular compartments. For these reasons I shall only discuss measurements made by the first method, TIRAF.

14.2 Theory of TIRAF for Cell Contacts

In order to provide a quantitative basis for TIRAF, Gingell et al. (1987) derived exact expressions for the electromagnetic fields in multilayered dielectrics, under several conditions of total reflection, in order to model the approach of a cell to an optical surface. Using these equations it is possible to express the cell-glass gap distance in terms of the relative fluorescence defined as the TIRAF ratio : (contact zone count rate/ background count rate). The refractive indices of the materials, membrane thickness, laser wavelength and angle of incidence of the laser beam at the glass/water interface (Fig. 1) must be known. Approximations valid for two of the cases that we analyzed have since been derived using different methods by Reichert and Truskey (1990).

We subsequently obtained experimental results in perfect correspondence with the theory (Mellor et al. 1988). The critical test employed thin insoluble films of MgF_2 vacuum deposited on glass. These are impermeable to fluoresceinated dextran, and act to prevent the fluorochrome molecules diffusing close to the glass. The refractive index of MgF_2 (n = 1.38) is similar to water (n = 1.33) so that the evanescent wave generated at the MgF_2/glass interface is almost the same as that at the water/glass interface. The result is that, where a MgF_2 film is present, the evanescent wave can only stimulate fluorescence at distances exceeding the MgF_2 film thickness.

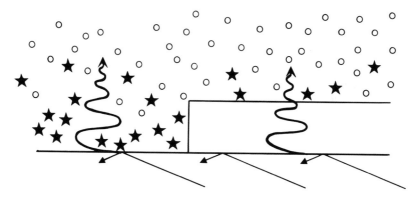

Fig. 7. Check of TIRAF theory using MgF_2 films which exclude dissolved fluorochrome from the region of high E^2 intensity near the glass. ★ fluorescent molecule, O non-fluorescent molecule.

This is shown in Fig. 7. Consequently, the fluorescence in the presence of the film is less than the background where no film exists, and fluorescence falls with increasing film thickness. Using a series of MgF_2 films, independently measured by a crystal oscillator method, we obtained the correct thicknesses by TIRAF, from 20 Å to 1600 Å (Fig. 8). This was achieved using several incident angles and a range of refractive indices in the aqueous medium from 1.339 to 1.448 ($\lambda = 488$ nm) showing that the excellent fit was not a happy coincidence under one particular set of conditions. The result proves that the theory correctly describes the behaviour of plane waves under conditions of total reflection in parallel multilayer films. Figure 9 shows the predicted and experimentally measured variation in TIRAF fluorescence with angle at a simple glass/water interface. The experimental values (•) fall satisfactorily on the theoretical curve.

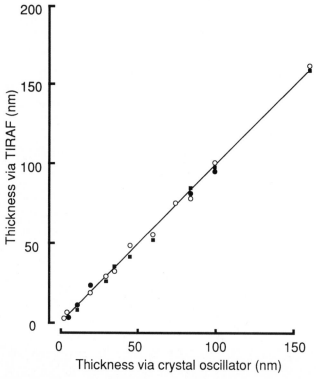

Fig. 8. Thickness of MgF_2 calculated from TIRAF theory corresponds with independently measured thickness. (**O**) medium refractive index n = 1.395, $\bar{z}$ =108 nm; (n) n = 1.339, $\bar{z}$ = 108nm; (l) n = 1.448, $\bar{z}$ = 69nm; (Simplified from Mellor et al. 1988).

Computer modelling of cell contacts based on the multilayer theory (Fig. 10) has produced insights into the interpretation of TIRAF images of cells. For example, the relationship between relative fluorescence and the dimensions of the aqueous gap beneath cells is rather insensitive to assumptions about the refractive index of the plasma membrane and the cytoplasm. Figure 4 shows that cytoplasmic heterogeneity does not intrude into the TIRAF image. This is particularly marked where a cell is spread

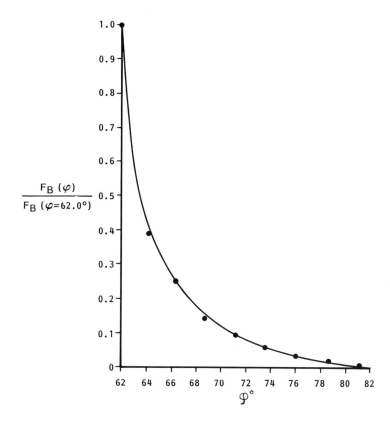

Fig. 9. TIRAF fluorescence versus incident angle, expressed as a fraction of the fluorescence at $\varphi = 62°$ close to the critical angle. (Gingell et al. 1987).

on glass of high refractive index (n_g = 1.85). On normal coverslip glass (n_g = 1.525) the results are sensitive to the second decimal place of the refractive index of cytoplasm, but virtually insensitive to the third. A crucial feature to emerge is that the relative fluorescence is a very sensitive function of the water gap in the region 0-100 nm. Roughly speaking, the evanescent wave is capable of 'probing' to a depth of 100 nm into the aqueous medium (i.e. water, cytoplasm or both). The depth decreases as the refractive index of the glass increases, and for a given value of the latter, depth falls as the angle of incidence is increased. Because of these features, TIRAF can be conveniently used to make a topographic map of cell-glass contacts in the range 0-100 nm. Optimal mapping sensitivity may require an appropriate choice of glass (refractive index) as well as measurements at a series of incident angles. Since the energy (proportional to the squared amplitude) of the evanescent wave falls exponentially, it is convenient to define a distance $3\bar{z}$ at which E^2 has fallen by 95%.

Depending on the sensitivity of the measuring equipment, this may be regarded as the practical cut-off point of the evanescent wave. On ordinary coverslip glass, cell-glass gaps exceeding $3\bar{z}$ = 200 nm cannot be measured, since $\bar{z}$ has a practical limit of around 100 nm, reached at an incident angle of 65° (the very rapid rise in $\bar{z}$ as φ_i approaches φ_{crit} obviates the use of higher angles). A moment's consideration shows that very small gaps are more sensitively measured using low $\bar{z}$ values.

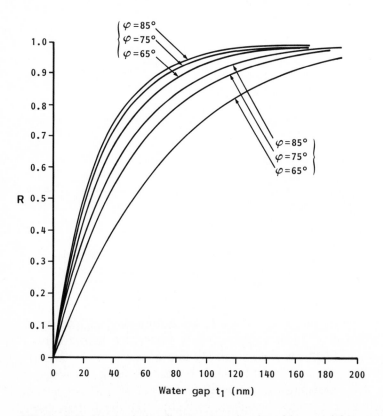

Fig.10. Calculated relative fluorescence developed in a variable aqueous gap t_1, (representing a cell-substratum apposition). *Three upper* curves, refractive index of glass n_1 = 1.85; *three lower curves*, n_1 = 1.539. Incident angles shown. Other refractive indices : aqueous medium n_2 = 1.337, membrane n_3 = 1.45 (thickness 4 nm), cytoplasm n_4 = 1.37, λ = 488 nm. (Gingell et al. 1987). Ordinate R is relative fluorescence, defined in text.

14.3 The Optical Effect of Thin Cytoplasmic Cellular Extensions

Many cell types develop lamellar extensions during spreading and locomotion on solid surfaces. Under IRM these can produce images which are very difficult to distinguish from dark zones of intimate contact, such as focal contacts, even when a quantitative study is made (Gingell 1981 ; Gingell et al. 1982b). A particularly convincing example of this is seen in amoebae of the slime mould *Dictyostelium discoideum* which have spread on a polylysine coated glass slides (Gingell and Vince 1982). Under IRM

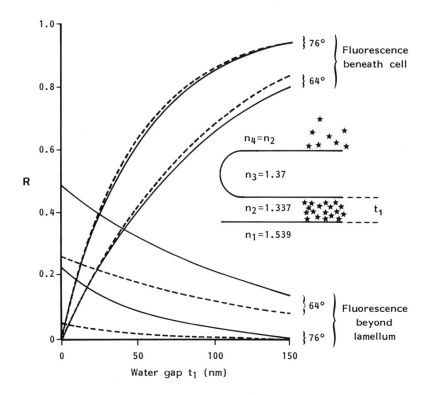

Fig. 11. Evanescent wave can penetrate a thin cytoplasmic lamella and stimulate fluorescence beyond it. *Inset sketch* shows details. *Ascending curves* : relative fluorescence beneath cell approaches unity as water gap t_1 increases. *Descending curves* : relative fluorescence generated beyond lamella falls to zero as t_1 increases. *Continuous curves* 100 nm thick lamella ; *broken curves* 200 nm lamella. Incident angles as shown. $\lambda = 488$ nm. (Gingell et al. 1987). Ordinate R is relative fluorescence.

optics, the cells rapidly form dark peripheral zones which develop centripetally. When viewed by TIRAF the contrast is reversed : the central region looks dark and the edges appear much paler (Todd et al. 1988 ; Fig. 3c, d). The reason for the appearance of the IRM image is that destructive interference across the ~0.1 μm thick lamella, makes it look black. Under TIRAF the central area is dark because the cell-glass contact there is too close to admit a significant amount of fluorescein. The edges are much less dark because the evanescent wave can penetrate the lamellar cytoplasm and still retain sufficient energy to excite fluorescence on the far side. This cannot happen in the central area where the cell is more than 1.0 μm thick. Figure 11 shows the way in which the fluorescence is predicted to depend on the thickness of the lamella, the penetration depth of the evanescent wave, and the aqueous gap under the cell. When the penetration depth is high ($\bar{z}$ = 100 nm) the lamella can show up to half the fluorescence intensity of the background, depending on the size of the water gap underneath. The analysis has been confirmed by electron microscopy which shows that the cell is shaped like a fried egg, with a "yolk" centre and a "white" periphery of the predicted dimensions (Todd et al. 1988).

These considerations show how a lamella behaves under TIRAF optics, but do not directly indicate whether the depth of the water gap under the centre of the cell is the same as that beneath the lamella. In other words, the two zones may have quite different contacts with the glass. This question was resolved by making TIRAF observations on high refractive index glass where $\bar{z}$ is low. In our study (Todd et al. 1988) we made observations on glass of index 1.83 at 488 nm, where $\bar{z} = 35$ nm. Under these conditions the evanescent wave has decayed by 94% at the far side of a 100 nm lamella ($e^{-100/35} = 0.057$, i.e. 5.7% of the energy at z = 0 remains at 100 nm). The images obtained by TIRAF show uniform darkness over the entire cell and no trace of the distinction between lamella and central region is seen (Fig. 12). This proved that the cell-substratum aqueous contact zone is uniform under the entire cell. This was a very satisfying result, since there is no way that such a deduction could possibly have been made from the complex IRM image, even using a proven quantitative theory of IRM image formation (Gingell and Todd 1979; Gingell et al. 1982a).

With the development of multilayer TIRAF theory and the realization that TIRAF offered a way of visualizing cell-substratum contacts, even when the thickness of some regions of the cell is extremely small, it seemed that the way was clear for making quantitative maps of cell contact topography. We were disappointed to obtain inconsistent results for water gap thicknesses beneath spread monocytes when measurements were made at several incident angles (Gingell and Mellor, unpublished). We even got variable results at a single incident angle if the degree of focus of the laser

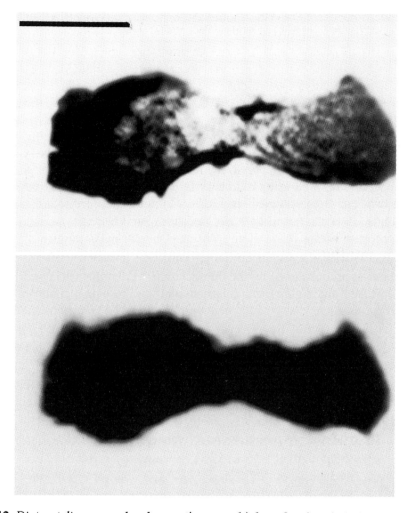

Fig. 12. *Dictyostelium* amoeba locomoting on high refractive index glass ($n = 1.83$; $\bar{z} = 35$ nm). The upper IRM image shows a dark lamellar region (*on the left*). None of the IRM complexities appear in the TIRAF image (below), which indicates a highly uniform contact over the entire cell. Digital image enhancement, from Fig. 4 of Todd et al. (1988). Scale bar = 10 μm.

beam was changed. The source of the error was tentatively put down to scattered laser light generated at the glass/water and glass/cell interfaces, since blue scattered laser light was clearly visible on removing the barrier filter which blocks wavelengths below 530 nm. After puzzling over the problem for several years, I realized that it is possible to determine the fluorescence contributed by light scattered near the TIR interface. It also became clear how to measure the additional error introduced by glare in the microscope, and these techniques are summarized below (for a full account see Gingell and Heavens 1994).

14.4 Compensation for Optical Noise in the TIRAF Image

The key idea in assessing the degree of fluorescence from the bulk of the fluorescein solution caused by laser light scattered at the glass/water interface (as well as at cell contacts) is the following. First measure the count rate from scattered laser light (C_1) in the presence of buffer or other aqueous media, but without any fluorochrome. Then add fluorochrome and repeat to obtain (C_2). Both measurements are made through a laser line pass filter that transmits no fluorescence. The difference $C_1 - C_2$ gives the laser scatter absorbed by the fluorochrome. From this it is not difficult to calculate the fluorescent emission and the resulting count rate. This requires a knowledge of the quantum efficiency (Q) a geometric constant for the capture of fluorescence by the microscope objective (g), and the relative efficiencies of the photocathode of the photomultiplier tube for laser and fluorescence wavelengths (ε) and the laser line filter transmission (f_s). A measurement of the count rate (C_3) for the fluorescent background is also made, in the presence of a laser barrier filter, whose transmission is f_f in the fluorescence band. The expression for the fraction of the total flux per unit area due to scattering is

$$P = K\left(\frac{C_1 - C_2}{C_3}\right) \tag{4}$$

where the system constant K is

$$K = \frac{f_f \varepsilon Q g}{f_s} \tag{5}$$

An *upper limit* for the geometric factor g, which is the fractional solid angle subtended by the aperture of the objective, can be readily calculated as

$$g = \left(1 - \cos(\frac{\alpha}{2})\right) = 0.28 \tag{6}$$

where α for the water immersion 63x Zeiss objective of numerical aperture 1.2 is given by

$$\alpha = 2\sin^{-1}\left(\frac{1.2}{n_w}\right) \tag{7}$$

where n_w for water at $\lambda = 488$ is equal to 1.339.

Using a measured value $Q = 0.53$ we find that $K = 0.29$. This maximum value compares with K=0.17 obtained from direct measurements on small scratches (Gingell and Heavens, 1994). For a simple background region without any object detail we find $P < 2\%$, indicating that scattered background light leads to a negligible error. When similar measurements are made on adherent spread blood platelets we obtain $P = 5\%$, which is still reasonably small. It should be emphasized that when such a preparation is observed under the microscope with the barrier filter (cautiously) removed, the scatter pattern at the locations of the cells is quite marked, so that without measuring P there could be no assurance that the ensuing error in fluorescence measurements would be small.

Glare in the optical system is a second potential source of error which is quite distinct from scatter at the plane of focus. Light entering the objective may become scattered within the system by dust or imperfections at the lens/air surfaces as well as at glass/cement junctions and metal lens mounts. Multiple reflections between the objective and the glass preparation (and between the condenser and the lower surface of the glass slide in transmission microscopy) may also give rise to "badly behaved" light which contributes to glare. Light from beyond the field of view may also enter the objective at large angles. This cannot contribute to the regular image, but in the instrument some of this light may get converted into virtually non- directional glare, and some of this may enter the photomultiplier aperture (PMTA). This can easily happen in TIRF because the illuminated area greatly exceeds the field of view. A related minor error peculiar to TIRAF is caused by fluorescence photons derived from scatter in the focal plane (discussed above) entering the aperture at large angles to the optic axis, such that they can only contribute to glare.

The significance of off-axis photons can best be appreciated by reference to diagrams showing the paths of photons which can and cannot enter the PMTA (Fig. 13a, b). The important point is that photons originating in the evanescent zone at P_1 *beyond* the radius of the measuring

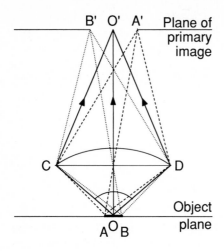

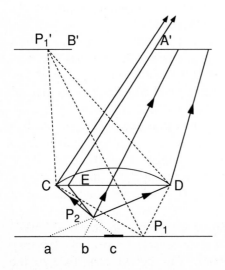

Fig. 13a, b. Paths of photons which can (**a**) and cannot (**b**) enter the photomultiplier aperture (PMTA) in a "well-behaved" way. The planar angle subtended by the lens at **O** in **a** is α of Equation 6. Area **AB** is imaged as a magnified conjugate area **A'B'** in the primary image plane. **A'B'** represents the PMTA and so all photons reaching the lens from **AB** pass into the photomultiplier. In the Zeiss photometric system there is a mirror at the primary image plane, with a small unsilvered spot forming the "hole" **A'B'**. The visible image is formed from points such as P_1 (**b**) reflected from the mirror and magnified by the eyepieces. The field is therefore seen with a black spot **A'B'** at the centre, since light striking this area passes unreflected to the photomultiplier. In **b** photons from P_2 have virtual origins a,b,c in the object plane.

spot AB (which is the conjugate image of the PMTA seen at the object plane in the Zeiss photometric system) cannot enter the PMTA. Referring to Fig. 13b, it is also readily seen that fluorescence photons generated in solution at P_2 by object scatter which have virtual origins beyond the radius of the measuring spot likewise cannot enter the PMTA directly. A proportion of them may, however, enter after suffering re-direction inside the microscope by any of the processes that lead to glare, as discussed above.

For TIRAF measurements we can now appreciate that it is necessary to make two sorts of glare measurements, Γ_2 for fluorescence photons in the presence of bulk fluorochrome under the conditions of measurement of C_3, and Γ_1 for laser scatter photons, under the conditions of measurement of C_1 and C_2. This makes it possible to correct scatter measurements for glare.

A convenient way of measuring Γ_2 involves the vacuum deposition of a 10 μm MgF_2 spot about 1000 nm thick, on a glass slide. This is so thick that no fluorescence can be generated in the fluorochrome solution on the far side of the spot. Consequently, when the (smaller) measuring area is brought into register with the MgF_2 spot, any counts registered will be due to glare. In the case of the glare factor Γ_1, it is convenient to make the appropriate measurement in an analogous way, using fluorescence rather than laser scatter. A monolayer of fluorescent protein is deposited on glass, the excess is washed off, and a spot several μm in diameter is photobleached in it using epi-illumination. The monolayer is then excited by TIR and any fluorescence photons measured from the bleached zone, in excess of the background value, represent glare.

Since glare distributes light more or less evenly across the field of view (Pillar 1977), and is proportional to the total flux through the optical system, we may write

$$F_G = F + \Gamma F \tag{8}$$

where F_G represents flux in the presence of glare, F is the 'ideal' flux and Γ is a proportionality factor equal to Γ_1 or Γ_2 as described.

We are now in a position to see how the correction for object scatter can itself be corrected for glare in measurements on cells. In the absence of stray light from any source, the ratio from which the cell-glass distance can be computed is

$$R = \frac{C^{iC}}{C^B} \tag{9}$$

where C^C is the count rate at a cell detail (usually an area ~ 1.0 μm dia.) and C^B is the corresponding background count rate. In TIRAF measurement of cells, $R \leq 1$. In the presence of scatter in the focal plane the expression becomes

$$R = \frac{C_3^C - K(C_1^C - C_2^C)}{C_3^B - K(C_1^B - C_2^B)} \qquad (10)$$

where superscripts C and B refer to object and background regions respectively. Where glare needs to be taken into account this becomes

$$R = \frac{C_3^C - \left(\dfrac{\Gamma_2}{1+\Gamma_2}\right)C_3^B - K\left[C_1^C - C_2^C - \left(\dfrac{\Gamma_1}{1+\Gamma_1}\right)C_1^B - C_2^B\right]}{\dfrac{C_3^B}{1+\Gamma_2} - \dfrac{K}{1+\Gamma_1}\left(C_1^B - C_2^B\right)}$$

(11)

where Γ_1 and Γ_2 are the glare factors defined earlier. When scatter from the background (but not the cell) is negligible, the second term in the denominator of equations 10 and 11 disappears and equation 11 reduces to

$$R = \frac{C_3^C}{C_3^B}(1+\Gamma_2) - \Gamma_2 - \frac{(1+\Gamma_2)K}{C_3^B}\left[C_1^C - C_2^C - \left(\frac{\Gamma_1}{1+\Gamma_1}\right)\left(C_1^B - C_2^B\right)\right]$$

(12)

In order to obtain the cell-glass distance (t_1) from the value of R it is necessary to compute a family of curves $R(t_1)$ treating t_1 as the independent unknown. This requires values of the laser wavelength λ, the angle of incidence φ_i, the thickness of the cell membrane $(t_2 - t_1)$ and its refractive index n_3, the refractive index of the aqueous gap n_2 and that of the cytoplasm n_4. The computation is refined until these parameters, together with a certain input value of t_1, give a value of R sufficiently close to that obtained from the corrected fluorescence measurements. The computation, which is very straightforward, is carried out as follows. The quotient of equations 17 and 22 of Gingell et al. (1987) gives the expression for the fluorescence ratio (equation 9 above) in the case of a cell separated by a water gap of thickness t_1 from the glass surface

$$R(t_1) = \frac{\left(\gamma_1^2 + \beta_2^2\right)\left\{-a_1^2\left(1 - e^{2\beta_2 t}\right) + a_2^2\left(1 - e^{-2\beta_2 t}\right) + 4\beta_2 a_1 a_2 t_1\right\}}{2\beta_2\left(\gamma_1^2 a_3^2 + \beta_2^2 a_4^2\right)}$$

(13)

Where

$$a_1 \equiv \left(\beta_3^2 + \beta_2 \beta_4\right)\sinh \delta_1 + \beta_3\left(\beta_2 + \beta_4\right)\cosh \delta_1$$

(14)

$$a_2 \equiv -\left(\beta_3^2 - \beta_2 \beta_4\right)\sinh \delta_1 - \beta_3\left(\beta_4 - \beta_2\right)\cosh \delta_1$$

(15)

$$a_3 \equiv \beta_3\left(\beta_4 \sinh \beta_2 t_1 + \beta_2 \cosh \beta_2 t_1\right)\cosh \delta_1 + \left(\beta_2 \beta_4 \cosh \beta_2 t_1 + \beta_3^2 \sinh \beta_2 t_1\right)\sinh \delta_1$$

(16)

$$a_4 \equiv \beta_3\left(\beta_4 \cosh \beta_2 t_1 + \beta_2 \sinh \beta_2 t_1\right)\cosh \delta_1 + \left(\beta_2 \beta_4 \sinh \beta_2 t_1 + \beta_3^2 \cosh \beta_2 t_1\right)\sinh \delta_1$$

(17)

$$\delta_1 \equiv \beta_3\left(t_2 - t_1\right)$$

(18)

In the above equations

$$\gamma_1^2 = n_1^2 k_0^2 - k_z^2$$

(19)

$$\beta_2^2 = k_z^2 - n_2^2 k_0^2$$

(20)

$$\beta_3^2 = k_z^2 - n_3^2 k_0^2$$

(21)

$$\beta_4^{\,2} = k_z^{\,2} - n_4^{\,2}k_0^{\,2} \tag{22}$$

where $k_0 = 2\pi/\lambda_{vac}$ in which λ_{vac} represents the vacuum wavelength of laser light, and finally $k_z = n_1k_0 \sin\varphi_i$ in which φ_i is the angle of incidence of the laser beam on the TIR interface. In this model the thickness of the cell is assumed to be much more than the characteristic penetration depth of the evanescent wave, and no continuous waves are set up in the membrane or cytoplasm. The conditions under which the latter may occur are discussed in Gingell et al. (1987) and the interested reader is referred to Fig. 2 b,d of that paper. We have begun to apply this method of correction to platelets spread on glass. Results obtained to date show clearly that glare makes a significant contribution to the images of cell contacts. The error becomes more serious as the contact becomes darker and the ratio $R = C^C/C^B$ decreases, since a given transfer of flux from background to cell produces a progressively larger error as C^C falls.

We have recently begun to make measurements on human blood platelets allowed to spread on glass (n = 1.525 at 488nm) in 1:1 plasma / phosphate buffered saline using the equipment described by Gingell et al. (1985) with minor additions (Todd et al. 1988). The results show that correction for stray light in TIRAF is crucial. Values of $\Gamma_2 = 0.15$ and $\Gamma_1 = 0.10$ were obtained at an incident angle of $\varphi_i = 76°$ ($\bar{z} = 60$nm). The system constant for external scatter, K (equation 5) depends on the quantum efficiency Q in addition to the other parameters which can be measured (see Gingell and Heavens 1994 for details) Using $Q = 0.53$ and the measured values of C_1, C_2 and C_3, equation 12 gives a fluorescence ratio $R = 0.30$ for a region of close apposition to the glass. This transforms to a contact distance $t_1 = 18$nm. If no correction is made for stray light, the value of R from equation 9 gives $t_1 = 32$nm which represents a substantial overestimate. Further analysis shows that scatter from the TIR interface ($P < 2\%$: equation 4) makes a negligible contribution to the calculated distance and that scatter from cell contacts is less important than glare. The results are not strongly dependent on Q, since an increase of 40% from the value used changes t_1 by only 10%.

References

Axelrod D (1967) Cell-substrate contacts illuminated by total internal reflection fluorescence. J Cell Biol 89 : 144-145

Axelrod D, Burghardt TP, Thompson NL (1984) Total internal reflection fluorescence. Ann Rev Biophys Bioeng 13:247-268

Gingell D (1981) The interpretation of interference reflection images of spread cells : significant contributions from thin peripheral cytoplasm. J Cell Sci 49:237-247

Gingell D, Heavens OS (1994) Method of correction for stray light in total internal reflection fluorescence measurement of cell contacts. (Submitted to Biophys J)

Gingell D, Owens NF (1992) Adhesion of *Dictyostelium* amoebae to deposited Langmuir-Blodgett monolayers and derivatized solid surfaces. A study of the conditions which trigger a contact-mediated cytoplasmic contractile response. Biofouling 5:205-226

Gingell D, Todd I (1979) Interference reflection microscopy : a quantitative theory for image interpretation and its application to cell-substratum separation measurement. Biophys J 26:507-526

Gingell D, Vince S (1982) Substratum wettability and charge influence the spreading of *Dictyostelium* amoebae and the formation of ultrathin cytoplasmic lamellae. J Cell Sci 54:255-285

Gingell D, Todd I, Heavens OS (1982a) Quantitative interference microscopy : effect of microscope aperture. Optica Acta 9:901-908

Gingell D, Todd I, Owens N (1982b) Interaction between intracellular vacuoles and the cell surface analyzed by finite aperture theory interference reflection microscopy. J Cell Sci 54:287-298

Gingell D, Todd I, Bailey J (1985) Topography of cell-glass apposition revealed by total internal reflection fluorescence of volume markers. J Cell Biol 100:1334-1338

Gingell D, Heavens OS, Mellor JS (1987) General electromagnetic theory of total internal reflection fluorescence : the quantitative basis for mapping cell-substratum topography. J Cell Sci 87:677-693

Harrick NJ (1967) Internal reflection spectroscopy. John Wiley & Co. Wiley-Interscience Publishers. New York

Heavens OS, Gingell D (1991). Film thickness measurement by frustrated total fluorescence. Optics and Laser Tech 23:175-180

Lanni F, Waggoner AS, Taylor DL (1985) Structural organization of interphase 3T3 fibroblasts studied by total internal reflection fluorescence microscopy. J Cell Biol 100:1091-1102

Mellor JS, Gingell D, Heavens OS (1988). Measurement of the thickness of deposited magnesium fluoride films by evanescent wave fluorescence : a critical test of the general TIRAF theory. J Modern Optics 35:623-628

Truskey GA, Burmeister JS, Grapa E, Reichert WM (1992). Total interference internal reflection fluorescence microscopy (TIRFM) 11. Topographical mapping of relative cell / substratum separation distances. J Cell Sci 103:401-499

Todd I, Mellor JS, Gingell D (1988). Mapping cell-glass contacts of *Dictyostelium* amoebae by total internal reflection aqueous fluorescence overcomes a basic ambiguity of interference reflection microscopy. J Cell Sci 89:107-114

Reichert WN, Truskey GA (1990). Total internal reflection fluorescence (TIRF) microscopy. 1. Modelling cell contact region fluorescence. J Cell Sci 96:219-230.

Weis RM, Balakrishnan K, Smith BA , McConnell HM (1982) Stimulation of fluorescence in a small contact region between rat basophil leukemia cells and planar lipid membrane targets by coherent evanescent radiation. J. Biol Chem 257:6440-6445

15. Electron Microscopical Analysis of Cell-Cell and Cell-Substrate Interactions : Use of Image Analysis, X-Ray Microanalysis and EFTEM

C. Foa, M. Soler, M. Fraterno, M. Passerel, J.L. Lavergne, J.M. Martin and P. Bongrand

15.1 Introduction

Numerous fundamental biological processes such as phagocytosis, lymphoid cell migration and T-lymphocyte cytotoxicity are triggered by cell adhesion. This phenomenon results from a series of complex interactions involving membrane and pericellular matrix molecules. The presence of polysaccharides on the cell surface is responsible for two distinct functional behavioral patterns : specific recognition phenomena and maintenance of a free intercellular space. Three non-exclusive mechanisms may play a role in cell-cell interaction :

First, the affinity between ligand structures may induce an accumulation of binding molecules into contact areas (Bell et al. 1984 ; Bell 1988).

Secondly, cell surface reorganization can be a consequence of some activation events, as shown in the interaction between T-lymphocytes and antigen-bearing cells (Kupfer et al. 1987 ; Kupfer and Singer 1989; Dustin and Springer 1989 ; André et al. 1990).

Thirdly, the movements of these adhesion molecules are complicated because the plasma membrane is not an isolated lipid bilayer with embedded proteins but has an associated pericellular matrix which may form steric barriers reducing the lateral mobility of membrane molecules (Zhang et al. 1991) as demonstrated with gold tagged lipids by Lee et al.(1993), and with proteins with mutated extracellular domains by Wier and Edidin (1988).

Moreover, there is growing evidence that membranes are made up of microdomains containing different groups of surface proteins. Thus it has been demonstrated that chemotactic receptors of neutrophil membranes are unequally partitioned between actin and fodrin rich microdomains (Jesaitis et al. 1988). A potentially important point, that recently received increased attention, is the influence of the cell coat on the adhesion process.

The glycocalyx or pericellular coat of non-adherent cells is not clearly delimited. Indeed the ectodomains of integral membrane proteins, proteoglycans and glycolipids intermingle with a variety of extracellular matrix glycoproteins and proteoglycans which contribute to a major part of cell surface charges and constitute a hydrated matrix of large size and structural complexity. In the literature, the glycocalyx, characterized by its sugar moities, is often opposed to structural glycoproteins such as selectins and integrins involved in the adhesion and recognition phenomena (Hynes 1992 ; Lawrence and Springer 1991). This matrix ranges in thickness from 10 nm to several micrometers depending on cell types. However, the structural and dynamic properties of the pericellular coat are now considered to play a central role in controlling cell adhesion. For example, cell coat thickness variations seem to regulate neural cell adhesion during embryogenesis (Yang et al. 1992)

In order to study these phenomena, it is important to obtain quantitative information on the structure of intercellular contact regions ; this was achieved by conventional fluorescence microscopy (Kupfer et al. 1986, 1987, 1989) and quantitative image analysis (McCloskey and Poo 1986 ; Ishihara et al. 1988 ; André et al. 1990). However, the apparent contact area observed with conventional optical microscopy includes regions that do not participate in adhesion. Indeed, electron microscopy allows easy detection of *electron microscopic free areas* where the intermembrane distance is incompatible with molecular bonding and *electron microscopic contact areas* that seem to be involved in attachment (Mege et al. 1987 ; Foa et al. 1988). Clearly, electron microscopy is the method of choice to study with optimal precision these morphological patterns :

- A computer-associated analysis of conventional samples may yield important information on the width of intercellular gaps, extension of contact areas, and adhesion-related reorganization of cell surface asperities (Mege et al . 1986, 1987 ; Foa et al. 1988).
- Immunoferritin or gold labelling of membrane molecules associated with statistical analysis (Hatson and Maggs 1990). This may yield additional information on molecular localization. This approach can also be applied to the study of the pericellular matrices using lectin conjugates.
- Histochemical quantitative electron microscopical study of pericellular matrix using X-ray microanalysis and Electron Filtering Transmission Electron Microscopy (EFTEM) may give accurate information on the elemental composition of well defined cell regions.

Therefore, quantitative cytochemical microscopic techniques have to be applied to study the pericellular coats during adhesion phenomena. This approach of the pericellular matrix distribution can be achieved by applying X-ray microanalysis to carbohydrates, sialic acid and proteoglycans stained with conventional reagents, since binding of cation dies is related to the amount of carbohydrates molecules. Thus we have applied it to ruthenium red stained material, as ruthenium is easily detected by this technique. For other staining procedures such as Alcian blue or lanthanum we have taken advantage of EFTEM both with spectroscopy (EELS) and imaging (ESI) to obtain semiquantitative determination of copper and lanthanum.

We shall now sequentially describe :

- the basic principles of sample processing
- selected methods of image analysis
- general procedure for quantitative elemental analysis.

15.2 Cells and Staining Procedure

Glutaraldehyde (2.5%) fixed bovine erythrocytes were allowed to adhere to murine macrophage-like cells (P388D1) by centrifugation in an Eppendorf tube containing 200 ml cacodylate buffer for 2 min. The supernatant was discarded and the fixative solution was carefully added to penetrate the pellet. For morphological observations and study by image analysis, we performed conventional fixation in 2.5 % glutaraldehyde followed by 1 % OsO4.

For the determination of the pericellular coat, three staining procedures were employed. We shall give only a brief description of these techniques and refer the interested reader to suitable references :

- *Ruthenium red* (Luft 1971). Fixation was achieved in a ruthenium red stock solution (0.05% at 60°C for 5 min ; after centrifugation at 1600 g for 10 min, keep the supernatant) (1 vol.) and 2.5 % glutaraldehyde in cacodylate buffer 0.2 M pH 7.4 (2 vol.) for 1 hour at 20° C ; rinse in the same buffer and postfixation for 3 hours in a mixture of 2% OsO4 (2 vol.) and ruthenium red stock solution (1 vol.).
- *Tannic acid with Alcian blue* (Montesano et al. 1984). Fixation was performed in 2.5% glutaraldehyde in 0.1 M cacodylate buffer pH 7.4 containing 0.5% Alcian Blue 8GX (Gurr Ltd, England). Then rinse in the same buffer and postfixation in 1% OsO4 in Veronal buffer for 20

min and after 1 min in 0.1% tannic acid in 0.05M cacodylate buffer pH 7.0 ; rinse in the same buffer and dehydration in graded series of alcohols.

- *Lanthanum* (Revel and Karnovsky 1967). Fixation was done in 2.5% glutaraldehyde in 0.1 M cacodylate buffer pH 7.4. Then rinse five times 20 min each in fixation buffer and twice for 10 min in HCL-S collidine buffer 0.2 M pH 7.4. Postfixation was achieved during 2 hours at 20° C in a mixture of 4% lanthanum nitrate (Sigma, USA) in HCL-S collidine buffer(1 vol.) and in 2% OsO_4 (1 vol.). After a very rapid dehydration in graded alcohols (3 min in each bath).

For all the techniques samples were embedded in Epon. Ultrathin sections were counterstained with uranyl acetate and lead citrate for morphological observations and left unstained for all the cytochemical observations.

15.3 Image Analysis

15.3.1 Purpose of the Method

As shown in Fig. 1, cell-cell contact areas display complex structural patterns that must be analyzed quantitatively in order to understand the binding process at the molecular level. Thus, the apparent contact area observed with conventional optical microscopy includes electron microscopic free areas where the intermembranar distance is incompatible with molecular bonding (EMFA) and electron microscopic contact areas that *seem* to be involved in attachment (EMCA).

We shall now describe general methods allowing quantitative study of the geometrical properties of cell-cell contact areas at the electron microscopical level. The basic idea is to digitize electron micrographs in order to extract numerical values for parameters likely to be related to the adhesive process. Since there is a growing variety of commercially available hardware and software that may be used for this purpose, our aim is to pinpoint the basic functions that must be available in order to solve problems related to adhesion. This should help the reader either to develop a "home made" system or to buy a suitable apparatus. The methods we shall describe have been incorporated into an image analysis software that has been developed in our laboratory and used for more than 5 years to analyze images obtained with fluorescence microscopy (André et al. 1990), conventional optical microscopy (Benkoel et al. 1990) or electron microscopy (Mege et al. 1986 ; Foa et al. 1988 ; Chamlian et al. 1991).

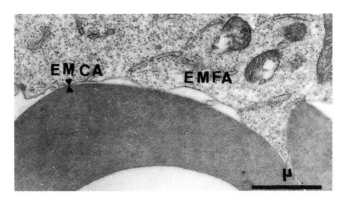

Fig. 1. Ultrastructural morphology of the interaction between glutaraldehyde treated erythrocytes and P388D1 macrophage-like cell. Bar=1μm.

15.3.2 Data Acquisition

The first step is to translate morphological patterns into a computer-usable form. Three basic procedures can be considered.

Digitizing pad

The simplest way of defining a cell boundary is to follow it manually with a digitizing pad. This procedure takes advantage of the investigator's capacity to recognize the cell surface and avoids sophisticated computing operations. According to our experience (Mege et al. 1986 ; Foa et al. 1988), a cell contour may be defined with about 1 mm accuracy, corresponding to an actual precision of 20 nm when a standard magnification of 50 000 is used. This is not enough, since this is the order of magnitude of the electron-light intermembrane gap found in contact areas. Also, when cell contours are not clear cut enough, this procedure is somewhat subjective, which may be felt unsatisfactory. These problems may now be overcome due to the increasing availability of new devices.

Optical Scanners

Many scanning devices can be acquired at a reasonable price. A standard apparatus allows a resolution of the order of 400 dpi (dots per inch : this represents a resolution of 0.06 mm) with 8-bit accuracy (i.e. 256 gray levels). This is more than sufficient, due to the usual quality of electron micrographs. Full page scanning devices can digitize an A4 format sheet without requiring any manipulation. Handscanners can digitize limited areas that have to be followed manually. This requires some care to avoid distorsions.

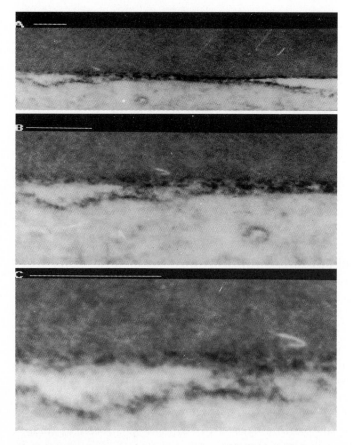

Fig. 2. A limited contact area between a glutaraldehyde treated erythrocyte and a P388D1 macrophage-like cell was scanned with three different magnifications from a 63000 enlarged electron micrograph (Bar= 0.1 μm). Images were obtained from computer files with a video processor. Cells were stained with ruthenium red.

Now we shall review some practical problems that must be considered in order to take advantage of this methodology. The first problem is that digitized files require a lot of computer memory. Thus, a 256-level image of 20×20 cm^2 size with a resolution of 400 dpi requires about

10 megabytes, which is higher that the memory of usually available computers. It is obviously quite difficult to store a high number of files for delayed processing. An easy way of dealing with this problem is to use hand scanners and digitize only limited areas corresponding to contact zones. See Fig.2.

There is another difficulty that is rapidly encountered, particularly with handscanners. Digitized images are obtained as coded files and there is a fairly high number of formats that may not be read by all image analysis systems. Sufficient information to decode a number of these formats is described in a recent book by Lepecq and Rimoux (1991). According to our experience, TIFF and BMP coding may always be used in the IBM compatible world (the importance of BMP is linked to its use in Microsoft's Windows). The PICT standard may be encountered in Apple's Macintosh computers. In some cases, data are compressed in order to save space. This must be accounted for by image processing systems. It must be emphasized that suitable transcoding software should be commercially available very rapidly.

Another difficulty may be encountered with some handscanners : Although digitization is performed with a number of gray levels, output files code binary (two-level) images : each point of the digitized image is represented as a dot matrix. The percent of white dots is proportional to the brightness of the corresponding picture element (or pixel) on the actual image. Thus, data processing requires some preliminary calculations to recover light intensities, with concomitant resolution loss.

An important point is to assess the significance of the number of gray levels expressed in a given image. It is important to emphasize that a paper photography is not an exact reproduction of an original image, and it is well known that contrast is highly dependent on the paper properties (see André et al. 1990, for a quantitative example). An example is shown in Fig. 3 : a PC Vision+ digitizer (Imaging Technology, Bedford, MA) was used to generate a 256x256 pixel image made of 256 parallel lines with brightness ranging from 0 to 255. This image was reproduced with a videoprocessor (Mitsubishi, Model P66E). It was then scanned with two Logitech hand scanners (model 32 and model 256 respectively) and TIFF files were decoded and used to determine local intensities. The relationship between initial and measured intensity is shown in Fig. 3B.

It is concluded that the numerical values of local intensities have no absolute meaning, since they are heavily dependent on the settings of measuring devices. However, quite precise relative values of these intensities can be obtained.

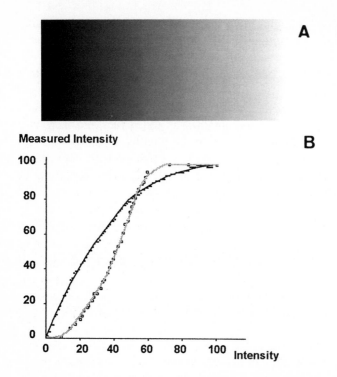

Fig. 3A, B. Limits of current commercial scanning devices. **A** a 256 level image made of 256 vertical lines with brightness ranging between 0 and 255 was constructed as described. **B** This image was scanned with two handscanners yielding 32 (*white circles*) and 256 (*black circles*) gray levels, and the measured intensity was plotted versus "expected" intensity, using a 100-level scale.

Using a camera and a digitizer

It is possible to digitize electron micrographs with a videocamera and a digitizer. Uniform illumination requires a convenient bench with a sufficient number of lights. The cost of this apparatus is substantially higher than that of a scanner. However, it may be useful to study rough surfaces (e.g. gels) that cannot be easily digitized with a scanner.

15.3.3 Contour Determination

The crucial step in studying a contact area is the determination of a linear structure, such as the boundary between the cell membrane and the extracellular medium or the cell membrane and underlying cytoplasmic area. The latter may be useful to measure the width of the intercellular gap, as will be described in section 15.2.4. In order to illustrate practical problems, we shall consider the cytosol-plasma membrane boundary displayed in Fig.2.

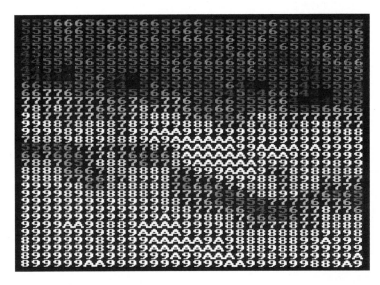

Fig. 4. Digitized image of the leftward zone of the contact area shown in Fig. 2A where each pixel brightness is represented with an hexadecimal sign.

The basic way of determining a contour is to chose a threshold value Vt and construct a line separating an area brighter than Vt from an area darker than Vt. The problem is that the choice of Vt plays a crucial role in contour determination. Obviously, the mere observation of Figs. 1 and 2 is of little help to make this choice. The following procedure has been used in our laboratory : a restricted area of the image is displayed on the monitor with each pixel represented as an hexadecimal values (0, 1, 2 ....,9, A, B, ...F) with a color chosen in order that the brightness match the intensity level. It is thus possible to recognize the region of interest and try a tentative threshold intensity Vt. As an example, the leftward zone of Fig. 2A is shown in Fig.4. Clearly, the brightness of the electron dense macrophage surface is of the order of 6-7. A quantitative analysis shows that the average brightness of the macrophage cytoplasmic region is 9.3 (standard deviation : 0.6) whereas the mean brightness of the intercellular electron-lucent gap is 9.7 (0.4 standard error). A reasonable order of magnitude for the threshold value between the macrophage cytoplasmic region and membrane is thus about eight intensity units. The influence of the choice of the threshold value is illustrated in Fig. 5A, B, C. These images were obtained by binarization of image 2A.

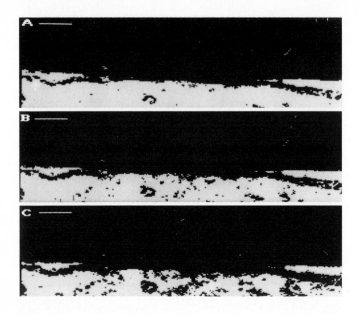

Fig. 5A-C. Binarization of the contact area shown in Fig. 2A with three threshold values : respectively 7.5 for **A**, 8 for **B** and 8.5 for **C**. Bar=0.1μm.

Pixels with brightness lower than the threshold were set black, and other pixels were represented with uniform brightness. The threshold values were 7.5, 8 and 8.5 in Fig. 5A, B, C respectively. Two frequent problems are clearly illustrated in these figures : if the threshold is too low (Fig. 5A), the dark zone corresponding to the macrophage membrane displays discontinuities (arrow), which prevents contour determination. If the threshold is too high (Fig. 5C) the contour will display artefactual asperities (arrow). A frequent problem is that the constrast of electron micrographs is often too low to permit automatic contour determination. Here are some methods that may be efficient in this case.

It is often possible to correct manually imperfect images. Thus, the continuity of the membrane shown in Fig. 5A (arrow) may be easily restored by joining the edges of the gap with a straight black line. In many cases, corrections are straightforward and there is no doubt about the location of the "true" boundary.

Some standard methods may allow a substantial improvement of images and facilitate contour determination (Toumazet 1987 ; Castleman 1979). The principle of median filtering is to replace the intensity value at any pixel by the median value of intensities of neighbouring pixels (e.g. the pixel + eight nearest neighbours in a square lattice). A straightforward modification consists of ordering the nine neighbouring intensities at any pixel and taking the ith value, where ith is a variable integer ranging between 1 and 9. This filter is easy to program, but it is relatively slow. However, a few seconds are required to filter a 256x256 pixel image with an AT or 386-based IBM-compatible computer with a short routine written with assembly language. An advantage of standard median filtering is that it may eliminate some noise without introducing any systematic error.

Another procedure is based on the enhancement of frontiers. Thus, the intensity of each pixel may be replaced with the difference between the highest and lowest intensities of neighbouring pixels. An example is shown in Fig. 6.

Fig.6. Same image as in Fig. 5 with described boundary-enhancement procedure. Bar=0.1µm.

According to our experience, the most satisfactory procedure is to use micrographs with reasonable contrast and accept the need of a few manual corrections. When this simple approach is inefficient, filtering techniques are seldom satisfactory, since boundaries may be too highly dependent on the precise method used.

15.3.4 Contour Analysis.

The analysis of digitized boundaries is usually much more straightforward than contour determination and can often be performed with simple BASIC programs. Since applications are quite diverse, we shall only describe some particular points of interest.

The most important quantity with respect to cell adhesion is probably the width of the intercellular gap in membrane areas. This is also the most tedious to determine : the only rigorous way to determine the distance between two digitized lines is to perform pixel-per-pixel calculation of the minimum distance between each point of a contour and all points of the neighbouring one. This may be done with a simple routine written with assembly language.

In many cases (see Fig.2) it is very difficult to detect a clear cut boundary between the intercellular gap and neighbouring membranes. In some cases, a useful trick was to determine first the mean distance between membrane-cytoplasmic interfaces of bound cells, and in a second step the mean membrane thickness of both cellular species. The latter values may be determined on free cell areas. The intercellular gap is then obtained by mere subtraction. Typical boundaries are shown on Fig. 7

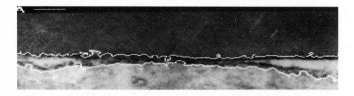

Fig. 7. Boundaries between intercellular gap of neighbouring membranes of the contact zone shown in Fig. 2A. Bar=0.1µm.

Another parameter of interest is the roughness of the plasma membrane. Clearly, cell adhesion requires substantial deformation of interacting surfaces at the submicrometer level. Quantitative methods were devised to estimate the cell surface rugosity in free and contact areas (Mège et al. 1986 ; Mège et al. 1987 ; Foa et al. 1988) and it was indeed demonstrated that at least one of adherent structures exhibited marked adaptation to the neighbouring surface. This adaptation probably requires active cell movements (Foa et al. 1988) and more work is required to model these deformations.

15.4 X-ray Microanalysis

15.4.1 Conditions of Observation and Data Acquisition

Unstained sections of approximatively 120 nm (gold interference color), mounted on copper grids, were stabilized with a carbon film (BALZERS 306 vacuum evaporator). The specimens were examined in a PHILIPS EM 400T electron microscope fitted with a 30 mm^2, 147 eV resolution lithium drifted energy dispersive X-ray detector. The standardization of analysis was obtained by switching off data acquisition when 6000 events reached the copper channel. The X-ray spectra were collected and analyzed with a TRACOR NORTHERN TN2000 multichannel analyzer. The microscope was operated in the STEM mode at 80 keV, 50 mA emission current, 25° specimen tilt, 45° take off and with a beam diameter of 50 nm. The spectra showed the presence of organic compounds but also revealed instrumental peaks (the largest one being due to copper at 8.04 keV, but silicium and chromium were also detected), from the specimen grid and holder with an associated extraneous continuum and a low energy noise due to electrons reaching the detector. Quantitative analysis was achieved by comparing Pic/Continuum as described by Hall and Gupta (1982).

15.4.2 X-ray Analysis on Pericellular Matrix Stained with Ruthenium Red. Semi -Quantitative Determination.

Pericellular coats of both macrophages and erythrocytes are stained by the divalent cation Ru^{++} which reacts mainly with glycosaminoglycans.
As seen on Fig. 8A, numerous erythrocytes adhere to macrophages in the conditions of the centrifugation (bar=1µm). Figure 8B shows clearly that the contact area both displays tight adhesion between cellular membranes but also loosely bound areas with large intermembrane spaces occupied by the pericellular coat. Molecular contacts seem to be distributed throughout the whole contact areas. Figure 8C shows a close contact area in which a ruthenium red stained material is accumulated. Figure 8D shows a focal site where the macrophage pericellular matrix is sharply defined. In contrast with Fig. 8D, Fig. 8E underlines the incidence of the plane of

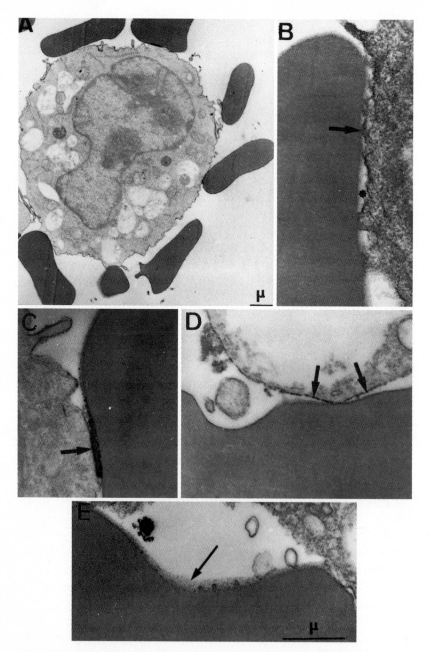

Fig.8A-E. The pericellular coats of macrophage and erythrocytes is stained by ruthenium red (**A**). Details of contacts areas are shown in **B**, **C** and **D** (*arrows*). **E** underlines the incidence of the plane of section on cell-coat thickness. Bar=1μm.

section on the thickness of the pericellular matrix. (Fig.8B, C, D and E. Bar= 0.5μm).

X-ray analysis was applied on pericellular matrix in free and bound areas. The peak/continuum ratios were calculated and compared in the free areas and in the bound areas. The obtained results do not fit with the local accumulation observed in Fig. 9A and 9C and raise many questions.

The most attractive hypothesis is that the pericellular matrix might contain microdomains with different behaviours during adhesion : pericellular matrix elements might be concentrated in some microdomains and rejected from others. These phenomena are demonstrated by the staining (Fig. 9A) and confirmed in ESI (data not shown) but they no longer appear when statistical averaging is performed due to the heterogeneity of the pericellular coat distribution. These results are consistent with the idea that the pericellular matrix density and thickness could modulate the intercellular adhesion.

Another problem is directly related to the artifacts of the staining procedure : we do not know if the ruthenium amount is proportional to the mass quantity of the matrix molecules (saturation ?) and whether the molecules contained in the contact area are fully accessible to the staining reagents.

15.5 Electron Energy Filtering Microscopy.

15.5.1 Principles

Energy Filtering Electron Microscopy (EFTEM) has been shown to provide elemental distribution within a cell with a high resolution (Ottensmeyer and Andrew 1980 ; Leapman 1986 ; Colliex et al. 1986 ; Colliex and Mory 1988 ; Martin et al. 1989 ; Horoyan et al. 1992). Electron energy loss spectra (EELS) and Electron Spectroscopic Images (ESI) were recorded on a ZEISS CEM 902 transmission electron microscope operated at 80 kV ; this EFTEM is equipped with a built-in Castaing-Henry magnetic filter which allows the selection of a given energy-loss range for the direct imaging of the specimen. The EELS spectra were recorded in the diffraction coupling mode, the scattering angle (17 mrad) was limited by the objective aperture ; the energy resolution was 1.5 eV (measured on the zero loss peak) and the specimen area to be analyzed was selected by the spectrometer entrance diaphragm ($7 \ 10^{-3} \ mm^2$). ESI was performed using the same angular selection ; the energy window in the final energy-filtered image was 10 eV, as limited by the adjustable energy selecting slit in the energy-dispersive plane behind the filter lens. For more information on the electron optics and working conditions with the ZEISS CEM 902, see for example Reimer et al. (1988). The microscope is equipped with a SIT low level camera and a KONTRON IBAS image analysis system which allows the recording and processing of the images. All the images were finally printed with a SONY UP 3000 color video printer. Careful attention was paid to avoid any modification of the gray scale during the recording time.

15.5.2 EFTEM on Lanthanum

Lanthanum stains muccopolysaccharides and muccopolysaccharide-protein complexes with high sensitivity. The electron opaque cellular coat is very similar to the ruthenium stained material and is very suitable for examination with EFTEM. Namely, lanthanum edges do not interfere with the edges of other elements. Lanthanum was evidenced using the $N_{4,5}$ edge located between 99 to 104.5 eV energy loss and the M_4 and M_5 edges around 830 and 849 eV. (Reimer et al. 1992)

Figure 9A clearly demonstrates the presence of lanthanum in the pericellular coat of a macrophage. The calculated background (in black) is printed for comparison. The characteristic $N_{4,5}$ edge of the element is very well defined and shows the shoulder characteristic of the element. Figure 9B shows the contrast of the staining at 0 eV energy loss resembling the aspect observed with conventional TEM. However, staining is sharply defined and reveals the heterogeneity of the pericellular coat. Figure 9C is an electron spectroscopic image performed at 130 eV energy loss corresponding to the $N_{4,5}$ edge of lanthanum. The heterogeneity of cellular coat distribution is underlined as the various brightnesses are directly correlated to the amount of the lanthanum staining. Figures 9D and 9E show respectively the 0 energy loss image and an image obtained by subtraction of two images taken at 120 and 108 eV energy loss.

As the confirmation for the presence of lanthanum in the pericellular coat, we have conducted some measurements using the M_4 and M_5 edges. The advantages of these two edges is that the two peaks are very well defined and this allows a higher signal to noise ratio (SNR) to be obtained. A spectrum is shown in Fig. 10.

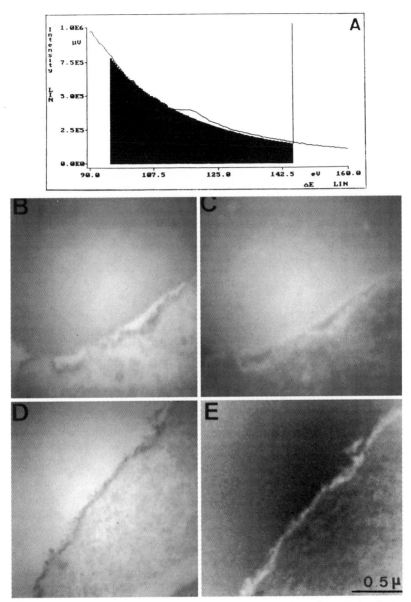

Fig.9A-E. EFTEM on lanthanum staining pericellular coat of a macrophage using the N$_{4,5}$ edge. **A** EELS of the N$_{4,5}$ edge of lanthanum (*arrow*) whereas the spectrum obtained just beside (limited by the dark area) does not show the shoulder characteristic of the element. **B, C, D, E** ESI of lanthanum ; **B** shows the contrast of the staining at 0 eV energy loss. **C** is an electron spectroscopic image performed at 130 eV energy loss corresponding to the N$_{4,5}$ edge of lanthanum. **D** and **E** show respectively the zero energy loss image and an image obtained by subtraction of two images taken at 120 and 108 eV energy loss.

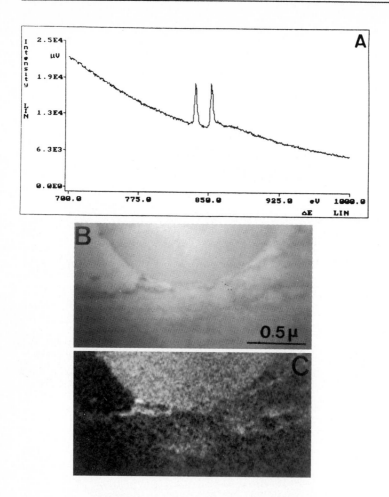

Fig.10A-C. EFTEM on lanthanum stained pericellular coat of a macrophage using the M4 and M5 edges. **A** EELS of the M4 and M5 edges of lanthanum. **B**, **C** ESI of lanthanum ; **B** shows the contrast of the staining at 0 eV energy loss. **C** is an electron spectroscopic image obtained by subtraction of two images taken at 850 and 840 eV energy loss.

15.5.3 EFTEM Applied on Alcian Blue and Tannic Acid Stained Cells

The cationic dye Alcian blue combined with tannic acid fixation does not specifically stain glycosaminoglycans, but protein carboxyl groups and other negative charges participate in dye binding. The pericellular coat of both macrophages and erythrocytes appears generally thicker than with the two aforementioned staining procedure and fibrils of about 250 nm are

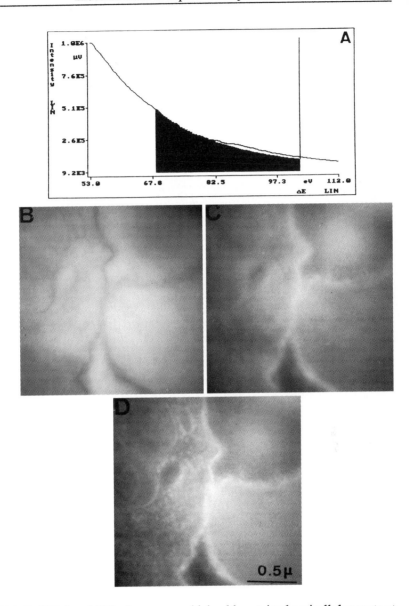

Fig.11A-D. EELS and ESI of copper on Alcian blue stained pericellular coats at the contact area of a macrophage and an erythrocyte. **A** Spectra obtained on the coat and in the cytoplasm. At 74 eV the shoulder corresponding to the copper $M_{2,3}$ edge is detected on the stained coat. **B, C D** : ESI on copper. **A** : 0 eV energy loss ; **B** : 110 eV energy loss which corresponds to the copper $M_{2,3}$ edge. **C** : 180 eV energy loss where the brightness is enhanced due to the contribution of the phosphorus edge.

visualized. The copper contained in the dye can be easily detected in EFTEM and evidenced using the $M_{2,3}$ edge at 74 eV energy loss and the $L_{2,3}$ edge between 930 and 950 eV (Reimer et al. 1992).

Figure 11 shows spectra and ESI images obtained on this material. Spectra obtained on the coat and in the cytoplasm appeared in Fig. 11A. At 74 eV the shoulder corresponding to the copper $M_{2,3}$ edge is detected on the stained coat. Figure 11B, C and D show the staining at respectively 0 eV energy loss where the contrast is maximum, at 110 eV energy loss which corresponds to the brigthess of the copper (just after the $M_{2,3}$ edge) and at 180 eV energy loss. There, the brightness is enhanced due to the contribution of the phosphorus edge (situated around 130 eV energy loss). The phosphorus containing ribosomes which are not stained by Alcian blue are visible in the macrophage cytoplasm.

15.5.4 Limitations of Semiquantitative Analysis of the Pericellular Matrix

However, studies about this matrix and particularly its quantitative analysis are difficult, mainly due to the following points :

- It is difficult to preserve the pericellular matrix with conventional fixations procedures. Here, we used conventional fixed samples but cryofixed specimens could have been examined. However, these techniques very often damage cell morphology and cannot define accurately enough the pericellular coat.
- The pericellular matrix is surrounding the cell and as we used sections, its chemical composition is not independent of the plane of the section.
- We have used preembedding techniques but we do not know if the contact zones during adhesion are fully accessible to the fixation and staining procedures. Postembedding staining have also to be employed.
- We do not know if the cationic binding is fully proportional to the amount of molecules and whether any saturation phenomenon can occur
- EFTEM and particularly ESI, although promising, sometimes raise problems of interpretation (Leapman and Andrews 1991). It is sometimes difficult to be sure that the brightness of an image corresponds only to the presence of the element as selected by its energy loss in biological specimens (Horoyan et al.1994).

References

André P, Benoliel AM, Capo C, Foa C, Buferne M, Boyer C, Schmitt-Verhulst A M, Bongrand P (1990) Use of conjugates made between a cytolytic T cell clone and target cells to study the redistribution of membrane molecules in cell contact areas. J Cell Sci 97:335-347

Ahn CC, Krivanek OL (1983) In Warrendale PA (ed) EELS atlas. Ashurem facilities, Gatan Inc., USA

Bell G.I (1988) Models of cell adhesion involving specific binding. In Bongrand P (ed) Physical basis of cell-cell adhesion. CRC Press, Boca Raton, pp 227-258

Bell GI, Dembo M, Bongrand P (1984) Cell adhesion : competition between non specific repulsion and specific bonding. Biophys J 45:1051-1064

Benkoel L, Gullian JM, Bongrand P, Dalmasso C, Xerri L, Brisse J, Sastre B, Chamlian A (1990) Use of Image Analysis to histochemical study of glucose-6-phosphate inactivation by diethyl pyrocarbonate in normal liver. J Histochem Cytochem 38:1565-1569

Castleman KR (1979) Digital image processing. Prentice Hall , Englewood cliffs N.J.

Chamlian A, Benkoel L, Bongrand P, Gullian JM, Brisse J (1991) Quantitative histochemical study of glucose-6-phosphatase in periportal and perivenous human hepatocytes. Cell and Mol. Biol. 37:183-190

Colliex C, Mory C (1988) Spatial resolution limits in EELS spectroscopy and imaging. Proceeding of the 46 th annual meeting of electron microscopy society of America, Milwaukee, USA, p 522

Colliex C, Manoubi T, Krivanek O L (1986) EELS in the electron microscope : a review of present trends. J Electron Microsc 35:307-313

Dustin ML, Springer TA (1989) T-cell receptor cross-linking transiently stimulates adhesivness through LFA-1. Nature 341: 619-624

Foa C, Mege JL, Capo C, Benoliel AM, Galindo JR, .Bongrand P (1988) T-cell mediated cytolysis : analysis of killer and target deformability and deformation during conjugate formation. J Cell Sci 89:561-573

Hall TA, Gupta BL (1982) Quantification for the X-ray microanalysis of cryosections. J Microsc 126:333-345

Hatson WS, Maggs AF (1990) Evidence for membrane differentiation in polarised leukocytes : the distribution of surface antigens analysed with Ig-gold labelling. J Cell Sci 95:471-479

Horoyan M, Soler M, Benoliel AM, Fraterno M, Passerel M, Subra H, Martin JM, Bongrand P, Foa C (1992) Localization of calcium changes in stimulated rat mast cells. J Histochem Cytochem 40:51-63

Horoyan M, Soler M, Martin JM, Benoliel AM, Fraterno M, Passerel M, Katchburian E, Bongrand P, Foa C (1994) Contribution of energy-filtering TEM to the detection of calcium : application to mast cells. J. Microsc 173:211-218

Hynes RO (1992) Integrins : versatility modulation, and signalling in cell adhesion. Cell 69:11-25

Ishihara A, Hopfield B, Jacobson K (1988) Analysis of lateral redistribution of a monoclonal antibody complex plasma membrane glycoprotein which occurs during cell locomotion. J Cell Biol 106:329-343

Jesaitis AJ, Bokoch GM, Tolley J O, Allen RA (1988) Lateral segregation of neutrophil chemotactic receptors into actin and fodrin-rich plasma membrane microdomains depleted in guanyl nucleotide regulatory proteins. J Cell Biol 107:921-928

Kupfer A, Singer SJ (1989) Cell biology of cytotoxic and helper T-cell functions. Ann Rev Immunol 7:309-337

Kupfer A, Singer SJ, Dennert G (1986) On the mechanism of undirectional killing in mixtures of two cytoxic T lymphocytes. J Exp Med 163:489-498

Kupfer A, Singer SJ, Janeway, CA, Swain SL (1987) Coclustering of a CD4 (L3T4) molecule with the T-cell receptor is induced by specific direct interaction of helper T-cells and antigen presenting cells. Proc natl Acad Sci USA 84:5888-5892

Lawrence MB, Springer TA (1991) Leukocytes roll on a selectin at physiological flow rates : distinction from and prerequisite for adhesion through integrins. Cell 65:859-873

Leapman R (1986) Electron energy loss spectroscopy. J Electron Microsc Techn 4 95-102

Leapman RD, Andrews SB (1991) Biological electron energy loss spectroscopy : the present and the future. Microsc Microanal Microstructure 2:387-394

Lee GM, Zhang F, Ishihara A, McNeil CL, Jacobson KA (1993) Unconfined lateral diffusion and an estimate of pericellular matrix viscosity revealed by measuring the mobility of gold-tagged lipids. J Cell Biol 120:25-35

Lepecq C, Rimoux N (1991) Les fichiers graphiques sur ordinateur. 1-Fichier Bit-Map. Armand Colin, Paris

Luft JH (1971) Ruthenium red and violet : 1. Chemistry, purification, methods of use for electron microscopy and mechanism of action. Anat Rec 171:347-351

McCloskey MM, Poo MM (1986) Contact-induced redistribution of specific membrane components : local accumulation and development of adhesion. J Cell Biol 102 2185-2196

Martin JM, Mansot JL, Hallouis M (1989) Energy filtered electron microscopy. Ultramicroscopy 30:321-328

Mège JL, Capo C, Benoliel AM, Foa C, Galindo JR, Bongrand P (1986) Quantification of cell surface roughness : a method for studying cell mechanical and adhesive properties. J Theor Biol 119:147-160

Mège JL, Capo C, Benoliel AM, Bongrand P (1987) Use of cell contour analysis to evaluate the affinity between macrophages and glutaraldehyde-treated erythrocytes. Biophys J 52:177-186

Montesano R, Mossaz A, Ryser JE, Orci L, Vassalli P (1984) Leukocyte interleukins induced cultured endothelial cells to produce a highly organized glucosaminoglycan rich pericellular matrix. J Cell Biol 99:1706-1715

Ottensmeyer FP, Andrew JW (1980) High resolution microanalysis of biological specimens by electron energy loss spectroscopy and by electron spectroscopic imaging. J Ultrastr Res 72:336-348

Reimer L, Fromm I, Rennekamp R (1988) Operation modes of electron spectroscopic imaging and electron energy-loss spectroscopy in a transmission electron microscope. Ultramicroscopy 24:339-354

Reimer L, Zelke U, Moesch J, Schulze-Hillert ST, Ross-Messemer M, Probst W, Weimer E (1992) EEL Spectroscopy : a reference handbook of standard data for identification and interpretation of electron energy loss spectra and for generation of electron microscopic images. Karl Zeiss electron optic division, Oberkochen, Germany

Revel JP, Karnovsky MJ (1967) Hexagonal array of subunits in intercellular junctions of the mouse heart and liver. J Cell Biol 33:C12

Springer TA, Lasky LA (1991) Sticky sugars for selectins. Nature 349:196-197

Toumazet JJ (1987) Traitement de l'image sur microordinateur. Sybel, Paris

Wier M, Edidin M (1988) Constraint of the translational diffusion of a membrane glycoprotein by its external domain. Science 242:412-414

Yang P, Yin X, Rutishauser U (1992) Intercellular space is affected by the polysialic acid content of NCAM. J Cell Biol 116:1487-1496

Zhang F, Crisse B, Su B, Hou Y, Rose JK, Bothwell A, Jacobson K (1991) Lateral diffusion of membrane spanning and glycosyl phosphatidyl inositol linked proteins : towards establishing rules governing the lateral mobility of membrane molecules. J. Cell Biol 119:75-84

Glossary

Adhesin The surface molecules of prokaryotes involved in adhesion.

AFM Atomic Force Microscopy.

Antecubital on the anterior side of the forearm. It is an usual site for venepuncture.

Antibody Protein species synthesized by the immune system of an animal that has been stimulated by the encounter of a given **antigen**. Antibodies are able to bind specifically to the antigen. They are very useful reagents in biological laboratories.

Antigen Any substance that is able to induce the synthesis of **antibodies** when it is injected into an animal.

Atherosclerosis is a very common degenerative disease of blood vessels resulting in increased rigidity, decrease of luminal area, increased risk of thrombosis.

Avidin A protein obtained from incubated hen's eggs which binds biotin with a very high association constant.

Bimorph Large displacement piezolectric actuator formed by two bonded piezoelectric strips that change length in oposite directions with applied voltage. Conversely, it may act as a piezoelectric force sensor. The bending of the bimorph produces a charge which is proportional to the bending. Hence, a measurement of the charge allows the determination of the bending of the piezoelectric spring and thus the force.

Biotinylation consists of derivatising a given substance with biotin, which is a small molecule with a very high affinity for **avidin** (Kd is about 10^{-15} M). This property is often made use of in microscopy. Thus, biotinylated antibodies are easily made fluorescent by incubation with avidin derivatised with a fluorescent probe.

Blocker A reagent added to react with a functional group so that it is no longer available for reaction, i.e. it is blocked., e.g. low weight amino compounds may be used to block tosyl groups so that they will not react with proteins.

Blotting The transfer of a part of a load of a macromolecule (s) from one material (say poly-acrylamide gel) to another material (e.g. nitrocellulose paper) by capillarity or electophoresis so that a range of recognition tests can be applied to the second (or third etc) material. e.g. Western blotting, electroblotting.

Butt joint Where the two structures which adhere form a joint without overlap of the structures, as opposed to a lap joint.

CAM   Cell adhesion molecule (of the homotypic type) where each cell bears the same type (s) of CAM that interact with the same molecular species on the other cell, as opposed to **SAM**.

Cantilever spring Leaf spring.

CD antigens Cluster of differentiation antigens. A large set of molecules, mostly expressed on human cell surfaces and mostly protein or of glycoprotein nature, recognised by (in the main) the use of antibodies generated against cell surface determinants. More than 86 CD antigens are known, many of them are recognised as being important in cell interactions and cell signalling. Different cell types, especially in the white blood cell group, can be defined and recognised by their possession of unique sets of antigens, even though any particular antigen may be expressed on several cell types. Related molecules are found in other species of mammal.

CD4 This is a surface molecule defining a subpopulation of T helper lymphocytes (or thymus-derived lymphocytes). It plays a role in both cell adhesion and activation.

CD8 This is also an adhesion and activation molecule of T suppressor and cytotoxic lymphocytes. In peripheral blood, a given T lymphocyte bears either CD4 or CD8.

Chemotactic means "able to attract cells through specific receptors". These receptors can sense a gradient of a particular **chemoattractant** and make the cell move towards regions of increased concentration of this substance.

Convolution refers to a mathematical operation consisting of combining two functions, f and g, to obtain a third function, h, called the "convolution product" :

$$h(t) = \int f(x)\, g(t-x)\, dx.$$

Cortex This is the external region of an organ (e. g. thymus, lymph node, brain)

Couette flow is the flow between two coaxial cylinders with uniform rotatory motion. The velocity has zero radial component and it is dependent on the radial coordinate r according to :

$$V = A\, r + B/r$$

where A and B are constants.

Cytoplasm A cell is made of a **nucleus** surrounded by a region called **cytoplasm.** It is surrounded by the **plasma membrane** that may be considered as the cell boundary.

Cytosol   The **cytoplasm** (see this term) is made of organelles (mitochondria, Golgi apparatus, endoplasmic reticulum, etc) embedded in a microscopically clear region defined as the **cytosol.**

DLVO theory   The acronym stands for Deryagin, Landau, Verwey and Overbeek, who proposed a theory of colloid stability based on the interplay of electrostatic forces of repulsion and electrodynamic forces of attraction.

Echinonectin   An adhesion related molecule of sea urchin embryos of the SAM type.

Ectodomain   Proteins are often made of "domains", i.e. structurally and functionally autonomous regions formed by a particular conformation of a polypeptide segment. A membrane protein often possesses extracellular **ectodomains** linked to intracellular domains through a transmembrane region.

Electrodynamic   Forces arising from the interaction of atoms, molecules or assemblages of molecules by electromagnetic polarisation as opposed to electrostatic forces. Also known as London forces, dispersion forces and Kirkwood-Schumaker forces.

Embryogenesis   This is the set of events involved in the formation of the embryo.

EELS   Electron energy loss spectroscopy. This consists of determining the energy loss of electrons interacting with a given material sample. exposed to the electronic beam of an electron microscope. This loss is related to the energy levels of atomic species in the sample, and it may be used to detect elements of interest (e.g. calcium, sulphur , etc).

EMCA   Electron microscopical contact area. This is the portion of a cell membrane (displayed on an electron micrograph) that is close enough to another cell or surface to be involved in adhesion. The definition of this area is somewhat arbitrary. The critical separation distance may be on the order of several tens of nanometers.

EMFA   Electron microscopical free area. This is the portion of the cell membrane (displayed on an electron micrograph) that is too distant from other cells or surface to be involved in adhesion.

ESI   Electron spectroscopic imaging. This consists of using EELS (see this term) to build images of material samples that represent the local density of a given element.

Endothelial cells   are cells coating the blood vessel walls.

Epithelial cells   line body surfaces (e.g. skin).

Erythrocyte   Red blood cell.

Epon   A resin that is widely used for sample embedding in order to perform electron microscopy.

Facstar Plus   This is the name of a popular flow cytometer (commercialised by Becton & Dickinson).

Fibrin is a derivative of **fibrinogen** (a molecule found in normal blood) that forms a solid meshwork that is essential during the initiation of the coagulation process.

Glioma cells are tumor cells derived from glial cells (i.e. cells surrounding neurones in neural tissues).

Glycocalyx It is now recognised that cells are surrounded by an "atmosphere" of large **proteoglycans** or **glycosaminoglycans** (see these terms) that may not be easily discriminated from extracellular regions, extending outwards tenuously from the cell surface. This was called "glycocalyx", or "fuzzy coat", or "extracellular matrix".

Glycosaminoglycans are long unbranched carbohydrate polymers that are important components of the extracellular matrix. Often attached to protein to form **proteoglycans**. They include keratan suphate, hyaluronic acid, heparan suphate, heparin, dermatan sulphate and chondroitin sulphate.

Haematocrit This is an apparatus used to measure the volume occupied by cells in blood. This word is also used to refer to the result of this measure, i.e. the percent volume occupied by cells.

Haemostatic means "that stops bleeding".

Homotypic adhesion refers to adhesion between similar cells or molecules (as compared to heterotypic adhesion between different structures).

Hypodermic means below the dermal layer. Hypodermic injections are made deeper than intradermal ones, and they are made with longer and larger needles.

ICAM-1 (Intercellular adhesion molecule 1) is a ligand of LFA-l. It is an adhesion molecule expressed by numerous cell populations, including endothelial cells.

Immunoferritin It is a widely used procedure in electron microscopy. Ferritin is a macromolecule with a large iron-rich core. It may be linked to a given cellular antigen through specific antibodies, which allows direct visualisation with an electron microscope.

Integrin This is a family of adhesion-related molecules with important connections with cytoskeletal elements and a capacity to mediate some kind of cell activation. They are made of two polypeptide chains (alpha and beta). There are three types of beta chains, defining three classes of integrins. Beta 1 integrins are often receptors specific for elements of the extracellular matrix (SAMs), and beta 2 integrins are found only on white blood cells.

Interferons are biological substances that are secreted by cells, and make cells resistant to viral infection.

Intradermal   Location of injections made immediately below the skin surface, e.g. to test allergic reactions. These injections are done with very small needles.

Isopachyte   A line mapping regions of equal thickness.

Langmuir-Blodgett films   are monomolecular layers of poorly soluble molecules that are formed on the surface of a liquid either by deposition (they are deposited as solutions in a volatile solvent immiscible with the liquid) or adsorption from solution (if their solubility is sufficient). Films of amphipathic molecules (i.e. molecules with an hydrophobic and hydrophilic pole) such as fatty acids may be considered as "half membranes".

LFA-l (lymphocyte function associated 1) is an adhesion molecule included in the integrin family. It is a beta 2 integrin found on white blood cells. It binds to ICAM-l, ICAM-2 and ICAM-3 (for Intercellular Adhesion Molecule). ICAMs may be found on white blood cells and endothelial cells. They can mediate the adhesion of white blood cells to the blood vessel walls, which is an early step of inflammation.

Medulla   This is the inner part of an organ (e.g. thymus, Iymph node).

Microvilli   are small protrusions - typically 0.1 μm diameter and less than a micrometer in length - found on the surface of nucleated cells.

May Grunwald-Giemsa staining is a (standard) procedure used by haematologists to stain white blood cells in order to be able to discriminate between different populations such as lymphocytes, monocytes or granulocytes.

Monoclonal antibody   Antibody synthesized by a single **clone** of lymphocytes, i.e. by lymphocytes generated by proliferation of a single progenitor. Monoclonal antibodies are homogeneous, in contrast with **polyclonal antibodies**, that may be extracted from the serum of animals injected with the corresponding antigen.

Mucopolysaccharide   This is an obsolete name for **glycosaminoglycans** (see this word).

Murine   means "of or from a mouse" (from latin mus, muris).

Neutrophils, or **Neutrophil granulocytes**   represent the most abundant population of white blood cells. They are actively phagocytic, and they are considered as the first cellular line of defence against bacterial infection.

Peeling   The breaking of an adhesion at a confined zone which moves through the adhesion as opposed to the breaking by a instantaneous break over the whole structure.

Perfusion is an experimental procedure consisting of generating a blood or liquid flow through an isolated organ or possibly an apparatus such as a flow chamber.

Pericellular surrounding the cell, located in the cell periphery.

Piezoelectric   A piezoelectric body is a very useful tranducer. It will generate an electric field/potential if it is subjected to an exogeneous force, and it will alter its shape if it is subjected to an electric field.

Plasmalemmal   means "about or of the plasma membrane", i.e. the cell surface membrane.

PMMA (polymethylmethacrylate) is a low adhesivity polymer that is used, in particular, to make contact lenses.

Poiseuille This physicist studied fluid flow in cylindrical pipes. Under laminar regime (i.e. **Poiseuille flow**), the fluid velocity v is everywhere parallel to the pipe axis. At distance r from the axis of a pipe of radius a, v is k $(a^2 - r^2)$, where k is a constant. **Poiseuille shear** is the shear rate dv/dr under Poiseuille flow. **Poiseuille approximation** consists of representing the flow in a pipe of slowly varying section as the flow in a cylindrical pipe of similar section.

Proteoglycan   Macromolecule made of a protein core linked to a varying number of unbranched polysaccharide **(glycosaminoglycan)** chains. The molecular weight may be higher than several million. Proteoglycans are found in connective tissues. They are constituents of the glycocalyx.

Proteoheparan sulphate   A   proteoglycan containing   as   its glycosaminoglycan heparan sulphate whose constituent   n-acetyl glucosamine is often sulphated. Hence highly negatively charged. Syndecan is one.

Raster scanning   Usually refers to devices such as computer monitors or television sets where rectangular images are built by sweeping sequential lines on a screen (with an electron beam) and illuminating sequential points (i.e. picture elements or "pixels") in each line.

Renal means " of or from the kidney".

Retardation effects, see also electrodynamic forces, since electromagnetic interactions propagate at the speed of light the interactions are modified when the propagation time between two assemblages is sufficiently long to decrease the mutual polarisation. Thus retardation effects become appreciable as the distance between two assemblages increases.

Reynold's criterion This is an empirical way of assessing the stability of a laminar flow. **Reynold's number** R is calculated as $vd\rho/\mu$, where v and d are a velocity and length characteristic of the system, ρ is the medium volume mass and μ is the medium viscosity. Laminar flows are stable in a cylindrical pipe if R is lower than about 1,000-10,000, depending on the shape of the inlet. When R is much lower than unity, inertial effects are negligible, and the equations of fluid motion are made notably simpler.

Rheoscope This is an apparatus consisting of a cone close to a plate perpendicular to its axis. The fluid, e.g. cell suspension, is placed in the gap between cone and plate. This may be subjected to controlled rotation, and the relationship between hydrodynamic resistance and rotation velocity may be used to determine the fluid viscosity. This viscometer also allows direct visualisation of the fluid, especially when this contains particles.

SAM Surface-associated (cell) adhesion molecule. Proteins and glycoprotein bound by integrins, and thus in a heterotypic manner as opposed to **CAMs**, and involved in cell adhesion, e.g. fibronectin, laminin.

Scanning Electron Microscopy consists of imaging the surface of a suitable processed sample by scanning this surface with a narrow electron beam and studying the scattering of these electrons.

Selectin This term refers to a recently described family of adhesion molecules that may be found on the blood vessel walls, white blood cells and platelets. The three known members of this family are E-selectin, L-selectin and P-selectin.

SFM Surface force microscope.

STEM Scanning transmission electron microscopy. It is a particular mode of operation of an electron microscope that may be used to perform elemental analysis.

Stenosis is a decrease of the section of a blood vessel resulting in decreased blood flow. This may be a consequence of **atherosclerosis**.

STM Scanning tunnelling microscopy or microscope.

Stroma this is a histological term usually referring to the matrix or basic structure of an organ of cells and intercellular material. Thus, thymus is made of a stroma and lymphocytes. In the thymus the stroma feeds and 'educates' cells of the T lymphocyte lineage.

Subendothelium It is the basal structure where endothelial cells are bound. It is highly adherent to blood platelets, in contrast with the endothelial cell layer separating blood from subendothelium.

Syndecan A transmembrane heparan sulphate proteoglycan found especially on the surfaces of epithelial cells and lymphocytes involved possibly in mediating cell adhesion.

Tenascin is a protein constituent of the extracellular matrix. It is involved in neural development.

Thrombin is formed in blood by cleavage of **thrombinogen** when coagulation is initiated. Thrombin will then induce the transformation of fibrinogen into fibrin.

Transmission electron microscopy is the standard way of practising electron microscopy : the sample is exposed to an electron beam whose deflection results in the formation of an image quite similar to a conventional optical image (using a suitable detector). The resolution limit is far better since the electron wavelength is much shorter than that of photons.

Tunnelling is a typical quantum mechanical effect that may be described with an example. If conducting surfaces are separated by an insulating medium that is in principle inaccessible to electrons (their energy is not sufficient to allow them to penetrate), the probability that an electron will jump from the first surface to the other is not rigorously zero, but it is an exponentially decreasing function of the insulator thickness. Thus, the determination of the electric current between surfaces may provide a very sensitive way of measuring the distance between them when this is low.

Thy-1   This is a molecule found on the surface of T-lymphocytes (i.e. thymus-derived lymphocytes, an important subpopulation of lymphocytes) and in brain. This has long been used to recognise T lymphocytes.

Tissue factor   Unspecified extracts released from tissues incubated in media, often the active principles which may aid growth, attachment or other biological properties, which, when recognised, are often found to be proteins.

Umbilical refers to the umbilical cord allowing the foetus blood supply during pregnancy. The umbilical cord is cut at birth, and the large veins found in umbilical cords are often used to prepare endothelial cells for experimental purposes.

Vascular   concerning blood vessels.